Exploring the Universe

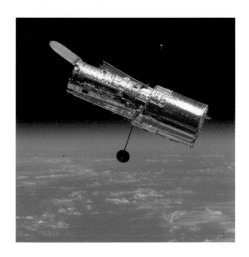

Exploring the Universe

THE ILLUSTRATED GUIDE TO COSMOLOGY

Brian Clegg

VIVAYS PUBLISHING

For Gillian, Rebecca and Chelsea

Published by Vivays Publishing Ltd
www.vivays-publishing.com

A catalogue record for this book is available from the British Library

ISBN 978-1-908126-16-0

Publishing Director: Lee Ripley
Design: Andrew Shoolbred
Illustrations: Jerry Fowler
Cover image: NASA/ESA/JPL-Caltech/UCLA/CXC/SAO, courtesy of nasaimages.org
Cover design: Tiziana Lardieri
Printed in China

Acknowledgements

Many thanks to Lee Ripley at Vivays Publishing for giving me the opportunity to write
this book. I would also like to thank Sir Patrick Moore for inspiring me as a teenager
to spend many chilly nights out with a small telescope, scanning the heavens.

Contents

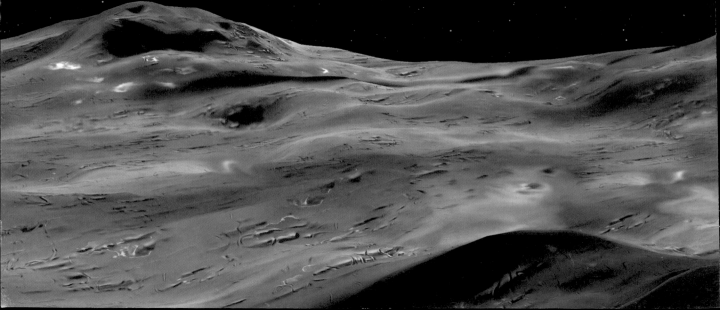

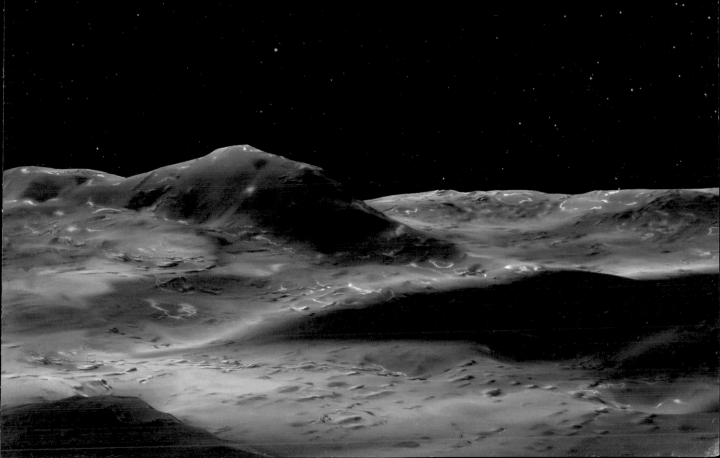

Introduction

There are few topics more awe-inspiring than cosmology. What is the universe? How does it work? Where did it come from? *Star Trek* got it right – space really is the final frontier, the last great unknown where explorers can make a host of new and exciting discoveries.

You might think that the exploration of the universe began on 4 October 1957, when Sputnik 1 was launched. Just 58.5 centimetres (under two feet) across, this fragile metallic ball sprouting two double antennae was humanity's first true venture into space. Although primarily a political gesture, this first artificial satellite did provide a small amount of scientific data.

Sputnik's 83-kilogram mass (51kg of that was its batteries) made waves totally out of proportion with its capabilities. Over the next 50 years we would see probes reaching the Moon, Mars and the outer solar system, a manned expedition to the Moon and a succession of manned space stations in orbit. Yet the most impressive explorers of all have been direct successors to Sputnik – unmanned satellites carrying instruments to tell us more about our universe.

Innovations like the Hubble Space Telescope and the COBE and WMAP satellites have been the true explorers of the universe on our behalf. In doing so, they carry on a tradition of visual exploration that goes back far further than any space flight – back past Galileo's telescope and Ptolemy's surveys of the sky to the very earliest humans.

When it comes to the universe, forget spaceships, light is our vehicle of choice. People have been exploring the universe this way ever since they looked up at the sky and wondered at the stars. With the naked eye it is possible to see the galaxy M31 in the constellation Andromeda. This appears as a faint smudgy spot in the sky, on the side of the constellation nearest the W of Cassiopeia. Telescopes have now shown that this smudge is a massive spiral galaxy, but even the unaided eye enabled early explorers of the skies to see it across 2.5 million light years – around 25,000,000,000,000,000,000 kilometres. Compare this with the furthest distance man has travelled – a mere 375,000 kilometres to the Moon.

What's more, light will always be our main means of exploring the universe. Light is the fastest thing in existence. If we could travel at half light-speed (something inconceivable with current technology because it would require a vast amount of energy), it would still take five *million* years to reach the Andromeda galaxy. We are never going to explore most of the universe in person, but light allows us to travel visually across immense distances.

We have come a long way since Galileo used his telescope to discover the first astronomical bodies not visible to the naked eye. We now use the whole spectrum of electromagnetic radiation – taking in radio, microwaves, infrared, X-rays and more – of which visible light is just a tiny segment. And with these remarkable engines of visual exploration we can venture out to experience the strange cast of characters that populate the universe: black holes and dark matter, supernovae and quasars.

With the information that has been gathered we can begin to formulate answers to the deepest questions about the universe and to speculate on its very nature and origins. This is exploration like no other.

chapter 1

Where did we come from?

Cosmology, my dictionary says, is the science of the universe as a unified object, plus the study of the laws that govern that whole. It's about the nature of the universe and how that universe came into being – some of the most fundamental questions of life.

This definition of cosmology assumes, of course, that we know what is meant by 'the universe.' The original Latin from which the word universe is derived means 'one turn', which doesn't help too much. But in practice it's fairly clear what we are dealing with. The universe is *everything* physical that exists from the smallest particle all the way up to the biggest galaxy. It is all matter, all energy, collected together as a whole as if this collection were an entity in its own right. That's an impressive concept, and it is natural to ask questions about it.

Earth and Moon, our local neighbourhood

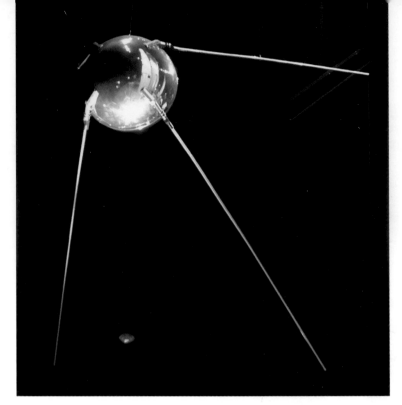

Replica Sputnik 1 at the National Museum of the United States Air Force

From the earliest times, creation myths have been written to explain where that 'everything' came from. Human beings are born storytellers. It's not our natural way just to pass on a series of dry facts, like a science lesson. Stories are more appealing, more memorable and a fundamental part of civilization. Creation myths are, in this sense, storytelling, not science. It's important to understand, though, that by calling these stories about the origin of the universe 'myths' we aren't insulting them, or those who consider them sacred.

In modern usage if we say of something 'That's just a myth', we are not particularly impressed with the idea. When applied to something like an urban myth, it's a commonly held belief that it isn't true. But the ancient myths were something entirely different. A myth was a story with a point. It gave information about a deep question, like 'Why are we here?' or 'Where did everything come from?' by reference to an exotic setting, usually distant in time. A myth was never intended to be the same thing as history – it didn't purport to be the absolute truth, but rather was a way of giving an understanding, a *feel* for today's reality, through a story of the past.

The early tellers of creation myths didn't have in mind the vast expanse of what we now think of as the universe. They could never have envisaged anything on that scale. For them, the universe was the Earth and the heavens (a rather vague term for everything that wasn't on the ground). Land, sea and sky accounted for all available space. Yes, there were a few oddities like the Sun, Moon and stars – but these were simply the inhabitants of sky, just as animals and people were the inhabitants of land.

To a modern eye, many creation myths can be confusing. It often seems that creation occurs in a strange order. So you might find that water is in existence before there is land (in which case, what holds it in place?), or light is created before any sources of light are made. But the vast majority of early creation myths have one thing in common – a creator. The answer to one of the biggest questions about the universe – how did it come into being? – was pretty well universally 'God (or 'A god') made it.'

This argument appealed to common sense. Many thousands of years later a Victorian clergyman called William Paley would use the same argument to explain how living creatures were formed. If you came across a watch lying on a beach, Paley said, you would not think it had naturally and randomly occurred. It was much too complicated and functional. Instead, you would assume that a watchmaker had created it. Similarly, when faced with the complex vastness of the Earth and heavens, the obvious response was 'It could only be like that if someone designed and made it.'

One of the earlier creation myths we know in detail is the ancient Egyptian description of the origins of the universe. To the modern eye, the Egyptian myths are highly confusing. Not only were there a multitude of versions, but also there was no definitive book or myth that was considered more real than any other. They did not restrict themselves to having a single god for a particular feature of nature (a Sun god, for example), and any particular god could have a number of aspects, where he or she looked, behaved and was even named differently. So the Sun was considered at various times

Paley argued the complexity of a watch implies a maker

to be different gods, or was aspects of a whole collection of gods.

Most versions of the first steps of creation in Egyptian myths came from water – something that is also seen in Genesis in the Bible. This seems to reflect the importance of the River Nile to Egyptian civilization. These aboriginal waters, called Nu or Nun, divided to provide dry land, on which was revealed the first of the gods, Atum. He spat out Shu, the god of the air, and Tefnut, the goddess of moisture whose daughter was Nut, the goddess of the sky and whose son was Geb, the Earth god.

It was these two siblings whose own children became the most important gods in the ancient Egyptian myths: Osiris, Isis, Set and Nephthys (though these were just a tiny fraction of the pantheon). In some variants of the creation myth, originating in the lower kingdom, the first of the gods was Ra, the Sun god, who was transformed into the Sun's disc, the Aten. In one short period of Egyptian history, this disc was thought of as the only god in the monotheistic religion of Akhenaten.

Early Chinese writings reflect the other common mythical starting point for the universe – an egg. In these myths, the creator god, P'an Ku emerged from an egg that seems to have just *been* since the beginning of time without ever being created. In the process of his hatching, the two halves of the egg became the sky and the Earth. But in this story, the true act of creation was one of sacrifice. At the age of 18,000, on his death, P'an Ku became everything that filled that egg-formed universe. His flesh became the soil and his blood formed the rivers. We, it seems, are descended from his fleas. In this myth, god and creation are all the same thing.

Probably the best-known creation myth in the West, though, comes in the first pages of Genesis, the opening book of the Bible. What we read here was developed from earlier Babylonian creation myths. The spirit of God moves over the face of the waters (we

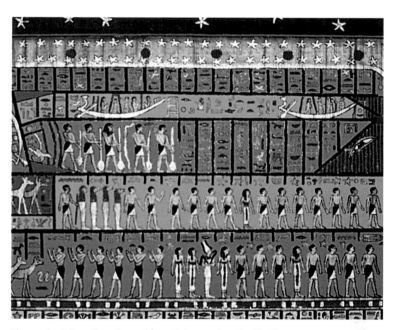

The ancient Egyptian sky goddess Nut spanning the Earth

The biblical Garden of Eden

aren't told where these waters come from, nor what is supporting them) and creates the Heaven and the Earth. To these are added light, plants, the heavenly bodies and living things, with man coming last.

Interestingly, immediately after this familiar creation story, the Bible then has an incompatible myth, if you think of these stories as history. In the Garden of Eden myth, man is created *before* the animals. But this is only inconsistent if you forget what myths really are. Each story has a function, the first one to establish God's role, the second to explain the human stewardship of the Earth and the nature of sin. There's no reason why they should be compatible. There is only a problem if you think that Genesis is talking about science or history – which it isn't.

Although many of the ideas in the Genesis creation myths go back much further, it probably came into its present form about the same time as another group of myths were being formulated – those of the Ancient Greeks. The Greek universe began with the void, emptiness – although confusingly they also referred to it as chaos, which seemed to imply some contents. The earliest god was usually Eros, born from the golden egg of the bird Nyx, an egg that split into two to form the sky and the Earth in a similar way to the earlier Chinese myths, although this version probably evolved independently. The idea of everything starting from an egg is a natural one – and when you see the universe as Earth and sky, it's not a great stretch to see these as two halves of a great egg.

In some versions – like the Egyptians, the Greeks had no definitive text – a mother goddess, Eurynome, was there alongside Eros and actively brought order out of the chaos. But the Greek gods we are probably most familiar with – from Zeus onwards (who himself would be the template for the Roman god Jupiter) – were the children of the Titans, a first generation of gods who were the children of Ouranos, the sky, and Gaia, the Earth.

The Greeks had their myths, like every other civilization that predated them, but uniquely in the story of our understanding of the universe, they went further. Rather than be satisfied with what could be called a magical explanation (the universe works because God makes it work), they looked for rational explanations of how the universe functioned.

chapter 2

Music of the spheres

Probably the first 'scientific' cosmology – a self-consistent picture of the universe and its origins that was built on physical forces and structures, rather than the whim of the gods – came from an early Greek philosopher called Anaximander. Born in Miletus in Anatolia (now part of Turkey) in the first half of the sixth century BC, Anaximander did not challenge the existence of the gods, but his view of the universe was based on simple observation rather than an attempt to get a mythological message across.

Unlike many of the creation myths that had the universe emerging from water, Anaximander preferred a beginning where everything came from an original chaos that was a sea of fire. This had one big advantage – it allowed him to explain a natural

Heavenly spheres - not to scale

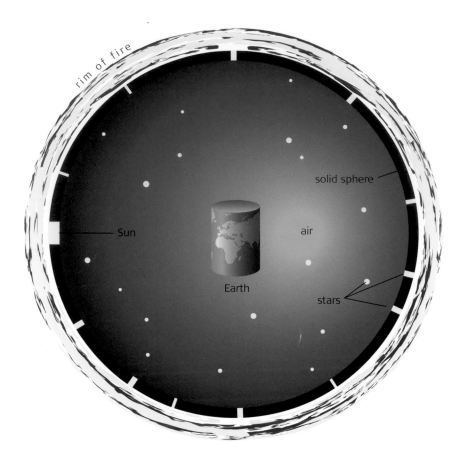

rim of fire

solid sphere

Sun

air

Earth

stars

Anaximander's sea of fire, shining through gaps as the Sun and stars

phenomenon. The early creation myths saw things arbitrarily brought into being by the gods, but Anaximander was looking for an explanation for the lights in the sky – the Sun, Moon and stars. He reckoned that this sea of fire still existed, and the universe was protected from the flames by a huge spherical shell. This shell had holes in it, and through these holes the fire escaped to provide the light of the heavenly bodies (and the heat of the Sun).

Anaximander and his contemporaries didn't give the universe much of a structure though – that would be the responsibility of the most famous of the Greek philosophers, Aristotle. The picture of the universe that Aristotle put together in the fourth century BC

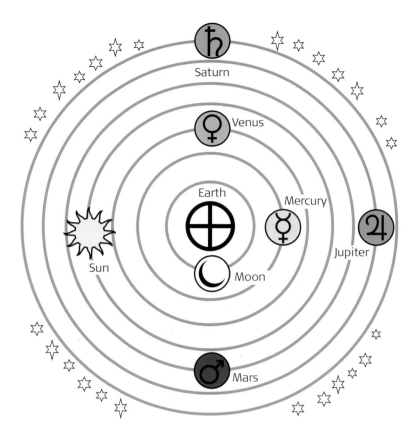

Aristotle kept Earth at the centre, but put the Sun into orbit

at Plato's Academy would become so rigidly accepted that it was, with some corrections to allow for contradictory observations, the model of the universe that remained in use for around two thousand years.

Aristotle put the Earth firmly at the centre of the universe, unmoving and immovable. This wasn't just a matter of human self-importance. His idea of how everything moved, from a dropped stone to a rising pillar of smoke, depended on heavy things being pulled towards the centre of the universe by a force called gravity, and light things flying away from the centre of the universe thanks to another force, levity. If the Earth were not the centre of everything, then you would expect heavy objects to fly off to some point in the sky that was the universal centre. They wouldn't stick to the Earth.

Around the Earth in Aristotle's picture were a number of crystal spheres, each nested within the other. In the first sphere was suspended the Moon, then Venus, then Mercury, then the Sun, followed by Mars, Jupiter and Saturn. Finally, came the sphere of the fixed stars. These weren't literally fixed in one place, in that their sphere was still thought to rotate, but they were fixed in the sense that they all moved together, while the planets (literally wandering stars) moved against them.

Gods still had a role in this picture. Each of the spheres drove the sphere within it, but something had to drive the outer sphere, the sphere of the stars, and this was put down to the 'prime mover', a deity. However this was still effectively a scientific cosmology – although a god was required to keep things moving, within the bounds of the universe everything functioned in a sort of heavenly clockwork.

In Aristotle's model, only the light from the stars could possibly come from outside the universe, but this wasn't how he saw things working. According to Aristotle, the Sun was the source of all light. Everything else – the Moon, the planets and even the stars were lit by reflected sunlight. When it was pointed out that you would expect the stars to be eclipsed, just as the Moon is eclipsed when the Earth gets between the Moon and the Sun, Aristotle argued that the Earth's shadow did not stretch beyond Mercury, so it couldn't eclipse the stars.

The universe that Aristotle believed we occupied was very small to modern eyes. It was little more than a rearranged solar system with the stars tacked on the outside. Yet it was still a massive place compared with Greece or any other part of the Earth, as another Greek philosopher Archimedes would discover when he set out to work out how many grains of sand it would take to fill the whole universe.

This sounds like one of those silly and meaningless exercises like trying to work out how many angels could dance on the head of a pin (something medieval philosophers are often said to have

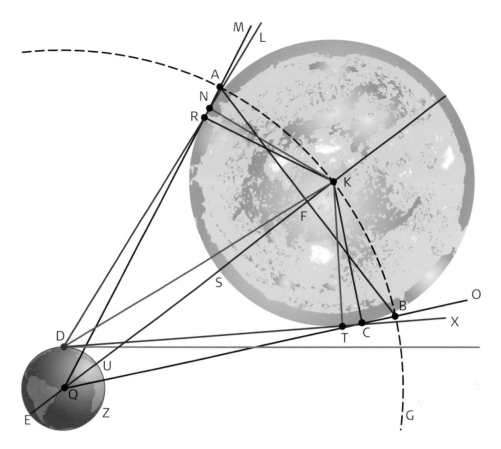

Archimedes' elegant geometry to work out the size of the universe

spent time worrying about, even though there is no evidence they did). But Archimedes was not engaged in worthless speculation – he had a serious intent in mind.

Archimedes was born around a hundred years after Aristotle, and was a much more practical philosopher. He indulged in sophisticated mathematics, coming close to inventing calculus, and designed a wide range of mechanical devices, from a screw to lift water out of the ground to giant curved mirrors that would have formed the first death ray had they ever been built, focussing the heat of the Sun on a wooden ship to set it on fire before it reached land.

In a little book called *The Sand-Reckoner*, Archimedes works out how many grains of sand it would take to fill the universe. Apart from being an entertaining exercise, his purpose seems to have been to illustrate how to extend the number system. Greek maths was limited because they biggest number they had was a myriad – 10,000. If you wanted to go really large you could have a myriad myriads, but that was it. Archimedes devised a number system that started from 100 million and built up to immense scales.

To work out how many grains of sand it would take to fill the universe, he had first to decide how big the universe was. Using a number of basic assumptions – like the Earth is bigger than the Moon, and the Sun is bigger than the Earth – and some fancy geometry, Archimedes was able to work out that the universe was around 10 billion stades across. This is a measure based on the size of a stadium, just as we often use football pitches as a measure for estimating distance today. Each of the stades was around 180 metres in length, so his universe was 1,800 million kilometres across.

We now know that 1,800 million kilometres is a little bigger than the orbit of Saturn, which isn't a bad estimate of the size of the solar system. In a tantalizing extra, Archimedes points out that the astronomer Aristarchus had brought out a book featuring the radical suggestion that the Earth moves around the Sun, rather than the Sun around the Earth. Unfortunately the book Aristarchus wrote on the subject has been lost, so this is the only known reference to the idea.

Because having the Sun at the centre would change his geometry, Archimedes reckoned that this version of the solar system would be about ten thousand times bigger, taking the diameter up to around 18,000,000,000,000 (million million) kilometres, which was more than enough to contain the main planets of the solar system. In the end, he reckoned it would take 10^{51} grains of sand to fill the ordinary universe – that's 1 with 51 zeroes after it – and 10^{63} grains to fill the Sun-centred version.

The idea Aristarchus had of putting the Sun at the centre of things seems to have been largely forgotten. It was Aristotle's model with an unmoving Earth around which everything rotated that would continue to be accepted until the sixteenth century – but relatively

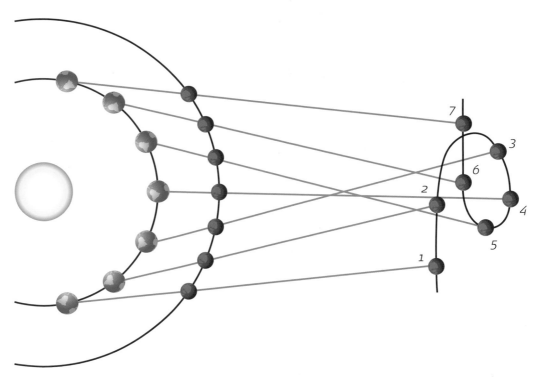

As Earth and Mars move round their orbits Mars seems to reverse in the sky

soon, the simple idea of everything turning in perfect spheres had to be modified to match reality.

This hadn't troubled Aristotle, because he was a philosopher who ranked thinking far above observation. So, for example, he said men had more teeth than women, but he never actually checked to see it was true. But the newer breed of philosophers who followed Aristotle felt it was important to check their theory against what could be observed – essential when moving from armchair philosophising to science. And the trouble was, some of the planets didn't behave themselves the way that the theory predicted.

If you plot the path that Mars takes through the sky, based on Aristotle's picture, you would expect Mars to follow a continuous path, making a circle around the Earth as it rotates on its crystal sphere. But, instead, you would find that Mars suddenly reverses itself in a process known as retrograde motion. It effectively performs a loop the loop in the sky, which didn't seem right and

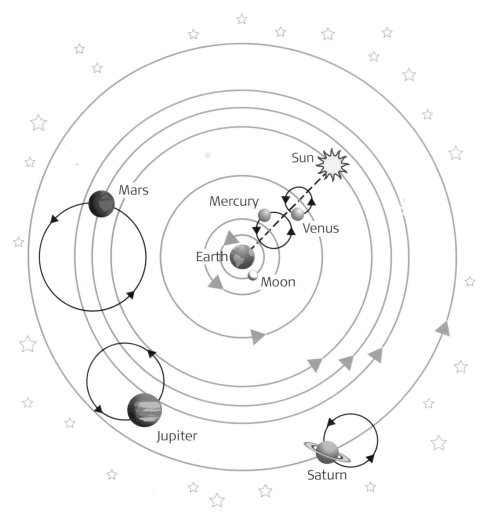

Ptolemy's epicycles, clumsily explained the reversing orbits of the planets

proper for something sedately moving on an unchangeable crystal sphere.

We now know that this strange motion is because Mars and the Earth are both rotating around the Sun, each travelling at different speeds on orbits that aren't concentric circles. So, as seen from Earth, the orbit of Mars will seem to loop back on itself as the faster-moving Earth overtakes it – but this wasn't a possible explanation when using Aristotle's model of the skies.

To explain this strange motion, from around Archimedes' time, it was suggested that a planet like Mars, instead of simply rotating around the Earth on its sphere also travelled in a separate circle called an epicycle. This was as if the sphere that held Mars had another, smaller sphere embedded in its surface, and this smaller sphere rotated too. So Mars would travel in circles around the little sphere as that little sphere moved around the Earth with the big sphere – producing the looping motion that was observed.

This version of the universe was given its most detailed description in the work of a later Greek astronomer, Ptolemy, who wrote a book in the second century AD that was later called *The Almagest*. (This is, confusingly, a Latinised version of an Arabic name for the work which roughly translates as *The Great Compilation*, although Ptolemy's own name for the book was *Mathematical Treatise*, and it was often known in Greek as *The Great Treatise*.) Ptolemy's version of epicycles, making the planets move in circles within circles, would form the definitive model of the universe right through to Galileo's time.

Galileo Galilei, born in 1564, wasn't the first in modern times to put the Sun at the centre of the universe. The Polish astronomer Nicolaus Copernicus had experimented with models putting the Sun at the centre to explain the complexity of planetary motion decades before. (His book on the subject was written in 1530, but he only allowed it to go into print on his deathbed in 1543 as he thought it would cause trouble – and he was right.)

Galileo's German contemporary, Johannes Kepler, took Copernicus' idea further. Kepler was a great believer in an ancient Greek concept, the music of the spheres. This was an idea that the

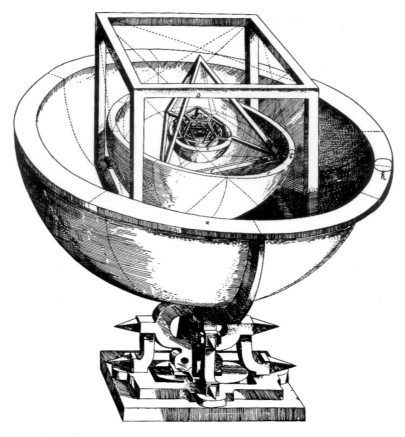

Kepler's vision of a solar system based on the platonic solids

movement of the spheres carrying the stars and planets produced a kind of heavenly music. This wasn't necessarily music you could hear however – certainly no one managed to pick it up in the quiet of the night. Instead this heavenly 'music' was a reflection of a special relationship that was thought to exist between the sizes of the spheres that housed the planets, just as strings on a musical instrument were harmonious when they were in certain ratios of length.

Kepler first tried to match up the imagined harmonious relationships of the heavenly spheres with the Platonic solids. These are the five basic three-dimensional shapes that can be made from collections of identical polygons. They range from the tetrahedron,

Kepler and Tycho Brahe contemplate the planetary orbits

made from four equilateral triangles, to the icosahedron, made from twenty such triangles. Kepler imagined each of the orbits of the planets being based on a sphere that fitted around one of the Platonic solids.

Although he never gave up on the idea of the music of the spheres, Kepler later realized that he could model the behaviour of the planets much better if he had them travelling around ellipses, squashed circles, rather than the perfect circles that the Greeks

Two of Galileo's early telescopes

(and Copernicus) had insisted on. He also found that if the planets moved in such a way that a line joining the planet to the Sun sweeps out the same area in equal times – so it travels faster when it is nearer the Sun in its elliptical path – he could match the best observations of the time on the movement of the planets, made by the Danish astronomer Tycho Brahe.

Galileo, famously put on trial for supporting Copernicus' system, added some powerful logic to support the idea that the Sun was at the centre of the universe, not the Earth. The ancient Greek model was totally founded on the idea that everything rotated around the Earth. Galileo didn't invent the telescope, but he did make an early one, and with it he began to study the heavens. He discovered that there were four moons orbiting around Jupiter. Here was direct evidence that everything did not rotate around the Earth, which called Aristotle and Ptolemy's models into question.

Galileo's punishment for defying the religious authorities could not hide the model that put the Sun at the centre of things. It just made too much sense, throwing away the need for those complex, messy epicycles. From the seventeenth century onwards, the picture of the solar system was pretty similar to our current one. At the centre was the Sun. Then came Mercury, Venus, the Earth (with the Moon orbiting it), Mars, Jupiter, and Saturn. We may have added a couple of extra planets, but in essence it was the same picture we now have.

A late portrait of Galileo

The stars were no longer thought to be on a crystal sphere, although this brought a new question: How did they manage to stay in place? And if the planets were just hanging in space, what kept them rotating around the Sun? Isaac Newton would put this down to gravity, a strange force that somehow acted at a distance to keep the planets (and us) in place – although it would take Albert Einstein to come up with an explanation of how gravity works.

Although detailed explanations would not arrive until the twentieth century, a new picture was emerging. The universe was opening up. Out there were planets, stars and more. But what are they, and where do they come from?

chapter 3

Building blocks

Our understanding of the universe has moved on considerably since Galileo's time. Taking a look at the universe through modern eyes we can see a sequence of building blocks that make up our understanding of the universe as a whole. We can zoom out from the most basic detail to provide all the essentials needed to put together a picture of the universe.

It starts at the most fundamental level with physics. All matter is made up of atoms. It was once thought that these were the absolute limit of our ability to divide matter up. That's why they are called atoms. The name comes from the Greek atomos, meaning uncuttable. Imagine cutting up something into smaller and smaller pieces. When you reach the limit and it is possible to cut no further, you have the most basic component, and this was called an atom.

Over time, gravity pulls matter in the accretion disk into planets

The idea of things being made up of atoms goes all the way back to ancient Greece, but it was a minority idea that was largely dismissed at the time in favour of a theory of everything being made of four elements: earth, air, fire and water. This dragged on for thousands of years (and is still used in some alternative medicine) but the concept of atoms would not go away. The idea came back as the early chemists began to investigate the chemical elements, finding that everything seemed to be made of a relatively small number of components. But most scientists still thought of atoms as an imaginary concept until the twentieth century.

We now know that atoms themselves are not the absolute limit for dividing up matter. Atoms are made of three kinds of particle – neutrons and protons in the massive nucleus and tiny electrons buzzing around outside in a cloud of uncertainty. You may well have seen illustrations of atoms where the electrons move in nice orbits like planets around the Sun, but science discarded this model nearly a hundred years ago. All we can say with certainty about electrons is that they are somewhere in a region around the atom, so they are best pictured as a fuzzy cloud surrounding the nucleus.

Of the particles that make up atoms, only electrons are as basic as matter can get. Neutrons and protons are made up of smaller particles called quarks, with three quarks per particle, though we never see quarks on their own because they are so strongly attracted to each other.

Add in a few particles that aren't directly involved in matter, most notably photons, which are the particles that make up a beam of light, and you have all the building blocks needed to construct a universe. But of themselves those blocks won't stick together. You also need forces – the mechanisms by which particles interact with each other.

There are four forces in total. Two of these are only involved within an atom – the strong nuclear force, which is what stops those quarks from being seen separately and keeps the nucleus together, and the weak nuclear force which is involved when an atom decays into different components. But the two forces

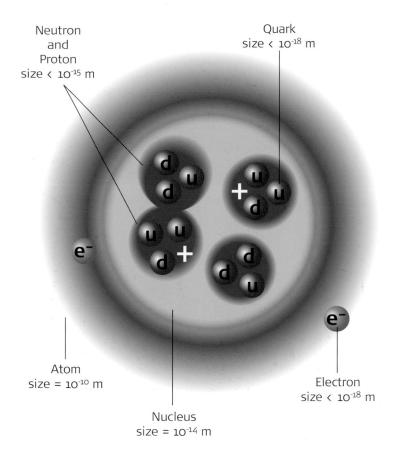

Neutron
and
Proton
size < 10^{-15} m

Quark
size < 10^{-18} m

Atom
size = 10^{-10} m

Electron
size < 10^{-18} m

Nucleus
size = 10^{-14} m

The inner structure of the atom: Electrons exist as clouds of probabality around the nucleus. To the scale shown for the nucleus, the whole atom would be around 10 kilometresaccross.

responsible for holding the universe together outside the atom are gravity and the electromagnetic force.

We are very familiar with gravity – and it will have a big part to play in the story of the bigger building blocks of the universe – but it is rivalled by the significance of the electromagnetic force. This is the force that holds the electrons and the nucleus of an atom together. It is also responsible for all the direct action between bits of matter. When you sit on a chair, your atoms don't actually touch the atoms of the chair. The electrical charges on the atoms in your body repel the charges in the atoms in the chair, rather like when

The electric charge on a comb overpowers the gravitational pull of the Earth

you bring two identical magnetic poles together. (The electromagnetic force is what makes magnets work too.) You don't sit on a chair – you float just above it.

One way that the electromagnetic force dwarfs gravity is that it is much, much stronger. If you rub a plastic pen or a comb in your hair for a little while it will acquire an electrical charge. Some of the negatively charged electrons in your hair rub off and stick to the plastic. Hold the charged object over a small piece of paper and the paper will jump up and stick to the surface, attracted by the electrical charge.

This ability to lift up a tiny piece of paper doesn't seem too remarkable until you consider what is really happening. The immensely massive Earth – around 6,000 million million million tonnes of it – is pulling down on that piece of paper with the force of gravity. In opposition, your little pen or comb is pulling up with the electromagnetic force of some electrons you rubbed off your hair. And the electromagnetism wins. The electromagnetic force from those few electrons is able to overcome the gravitational force of the whole Earth.

To make our universal building blocks we need matter, electromagnetism to make that matter stick together on the small scale, and gravity to work on the large scale. The universe is scattered with matter, mostly the gasses hydrogen and helium (we'll find out where they came from later). There is also a scattering of all the other elements up to uranium in the form of fine dust that is suspended through space.

Just imagine these clouds of matter floating around in space. There is no weather. There is not a touch of wind to move the matter around. But there is gravity, and, although the force between the atoms of gas and particles of dust is absolutely tiny, every last speck of matter is attracted by every other one. Those that are relatively near to one another will gradually, over aeons, move towards each other. Initially there are only tiny amounts of matter present in any particular part of space. But, over a vast amount of time, those fragments, of mostly hydrogen and helium, will begin to collect.

Once some matter has clumped together it will have a bigger gravitational pull and will drag in more of the gasses and space dust around it. If there is enough matter, it will start to produce a decidedly heavy object. All that matter is pressing in on itself. As more and more particles of matter crash into the object, their energy of movement, produced by gravity, is changed into heat. (Think of rubbing your hands together – the kinetic energy of movement gets changed into heat energy by friction.) Painfully slowly, the ever-growing ball will begin to heat up.

After several million years, a critical point will be reached. At this stage, three things are combining to make a remarkable reaction happen. The most common constituent of the ball, just as it is the most common substance in space, will be the simplest of the elements, hydrogen. The hydrogen atoms (or, more accurately, the hydrogen ions, which are hydrogen atoms with the electrons stripped off by the heat) will be pushed together under high pressure due to the gravitational attraction of a body that will now weigh several billion tonnes. The temperature at the core of this ever-expanding body will have soared. And something else remarkable will also be happening.

Particles like these hydrogen ions don't obey the rules we expect of matter on the scale of a person or a house. They are quantum particles, and rather than following ordinary mechanics, they obey quantum mechanics, the rules of which were discovered in the first half of the twentieth century. One of the special characteristics of quantum particles is that, until they interact with something, their exact location is uncertain. (This is why electrons in an atom can be explained as fuzzy clouds of probability, rather than travelling around clear circular orbits.) This uncertainty of location means that quantum particles can jump from one place to another without passing through the space in between, a process known as quantum tunnelling.

The positively charged hydrogen ions repel each other because of the electromagnetic force. Even under the temperatures and pressures that have built up, they can't get close enough to interact. But by tunnelling, a small percentage of those ions will jump to be too close to another ion. When they get so close, the strong nuclear force, which only operates over very short distances, takes over, attracting them together more strongly than the electromagnetic force repels them. Get past that limit and the hydrogen ions fuse to make a new type of ion – a helium ion, the next element in the periodic table.

In the process, a small amount of mass gets converted into energy. The equation that tells us how much energy we get when mass changes into energy is probably the best-known equation in all of history: $E=mc^2$, where E is the energy, m the mass, and c the speed of light. The speed of light is very big – so even a tiny amount of mass produces a huge amount of energy. If you could convert one kilogram of matter into energy, you would liberate the amount of energy that a typical power station produces in six years in an instant. Once the fusion process begins, there is a vast outpouring of energy. This blasts out in the form of electromagnetic radiation – light. A star has formed.

Stars are the most obvious building blocks of the universe. Take a look at the sky at night. Apart from the Moon and a few planets, all you will see is stars. With our most powerful telescopes we can make out billions of them. Unlike the planets and the Moon, which shine with reflected light, stars are nature's lamps. And they are responsible for much more, something that is fairly obvious when you realize that the Sun is a star. It's a pretty ordinary everyday star – but it's our own star and it is responsible for just about everything that keeps us alive.

If the only things that formed from those clouds of matter were stars, the universe would still be a dramatic place, but there would be no one around to see it. No form of life as we understand it

Hydrogen undergoes nuclear fusion in the Sun to produce helium

The gravitational
pull of a newly
formed star
drags the matter
around it into
a disk that will
coalesce into
planets

can exist on or in a star. But not all the matter in the vicinity of a star will be drawn into this vast nuclear furnace in space. There will usually be plenty of other matter left surrounding the star.

If the star were stationary, over time this matter would also fall into it. The more distant particles would take an extremely long time, but gravity's inexorable pull would gradually win and suck them in. But every star we've ever observed is spinning, and there is reason to think they all will be. This happens because the material they form from has some movement, most likely caused by nearby stellar explosions, which gets converted into spin. If you could start from a perfectly still area of space, you could form a star that didn't spin – but that isn't going to happen.

As the star forms and the matter pulls together, the spin speed of the material will increase, rather like when a skater speeds up if they pull their arms in while they are spinning round. The amount of oomph in the spin, a function of the rotation rate and how far away from the centre the mass is (called the angular momentum) stays constant – so if the object contracts, the spin has to get quicker to have the same amount of angular momentum.

The material around a spinning star, which is itself rotating, will end up like pizza dough spun between the hands – it will form a disc around the star called an accretion disc. Within this disc, similar processes to that which occurred during the star's formation will happen. Particles will be attracted together to make bigger and bigger forms, eventually making up planets. (In principle another star could form, and this often does, resulting in a binary system where two stars orbit each other.)

The parts of the disc around the star where more of the matter is made up of heavy atoms – attracted more strongly by gravity, so typically nearer the star – will tend to produce rocky planets like the Earth. Where there is less heavy matter, the composition of the planet will be more like the Sun (primarily gaseous) but without enough matter to reach the mass required to start fusion. The result will be a planet largely made up of gas, like Jupiter.

Although planets won't heat up anywhere near as much as a star, they will be warmed by the same process as new particles zap in, usually getting hot enough to produce a molten core, which may be kept in molten form (as is the case with the Earth) if there are enough radioactive elements in the planet to keep the heat flowing as the planet matures.

These, then, are the basic building blocks of a universe on the middle scale – planets and stars. For a long time this was thought to be as big as universal building blocks get, but since the eighteenth century there have been suspicions about some fuzzy patches in the sky. As we have developed better telescopes it was discovered that these little blobs were huge clusters of stars, each containing billions of suns. As we will discover they would become known as galaxies.

If weren't for the same spinning effect that keeps the matter from around a star from falling into it, there is no reason why all the stars in a galaxy would not gradually be pulled together into a single incredibly large lump by gravitational attraction. But, like stars, galaxies also are spinning, and this keeps them in a disc-like structure that often forms as a spiral, rather like a spiralling liquid heading down a plughole.

There are also collections of galaxies – cluster and superclusters – making up larger structures across the universe, so the addition of the galaxy has given us our basic building blocks to construct a universe. Stars at the heart of everything, planets orbiting around stars, and galaxies made up of collections of stars. Of course there are plenty of other inhabitants in the universal zoo – and we will come across them later, but we have now met the primary building blocks for a universe.

As we look into the night sky, if there are no street lights to dim the view, and if the Moon is not washing out the skyscape, we might see a few thousand stars. We know that there are billions upon billions more out there, but that is all our naked eyes can make out. This might not seem too surprising – eyes are just eyes, after all – but we shouldn't underrate them. Eyes are pretty special instruments.

The human eye is sensitive enough to detect a handful of the photons that make up light. We can see a candle flame around 16

Everything from the tree to the stars are made of the same atomic building blocks

kilometres (10 miles) away if the night is dark enough. In practice it rarely is as dark as this, but our eyes are impressively sensitive. Yet, given that sensitivity, there is something strange about the view we have when we stare out into the universe.

Think about it. Go out on a clear night and take a look at the sky. What do you see? Mostly blackness. Nothing. We are so used to this that it doesn't strike us as being odd. Black is what the night sky is – it just seems natural. Yet astronomers spotted something strange about that black night sky, something odd enough for it to be given a name of its own. It is referred to as Olbers' Paradox.

This problem had been thought of as long ago as the sixteenth century, but it was most widely publicized by the German astronomer Heinrich Olbers in the 1820s. It goes something like this: Let's assume that the universe is infinitely big – back then there was no evidence to suggest it wasn't. And we'll assume that stars are randomly scattered through space. Then if you pick any random line from Earth going up into the sky, you will eventually hit the surface of a star, which should be bright. So the whole sky should be a blaze of light, whichever way you look, rather than a black sheet with the occasional brilliant spot in it. The sky should be like one enormous star, day and night.

One of the first attempts to explain the blackness of the night sky was to suggest that there must be dark clouds of dust between us and many of the stars. Yet that was a problematic suggestion. The universe seems to have been around for a long time. Over the aeons, after constantly absorbing the starlight, those clouds of dust would heat up until they themselves began to glow. There still should be light in all directions.

Another thought is that perhaps the universe is not infinitely big. We really don't know how big the universe is, but it is entirely possible that it has a finite scale with plenty of room for gaps between the stars. Take it to the extreme – imagine the universe only consisted of the stars we could see with the naked eye. Then there would be lots of black sky.

This isn't a bad explanation, and it was one put forward by Edgar Allen Poe in an essay in which he also gave the explanation

Edgar Allen Poe was the first to realize the speed of light would limit the light reaching us from the stars

that is largely accepted now, only to dismiss it. Poe said that if the universe was infinite, or just extremely large, then the only acceptable reason for the blackness is that there hasn't been enough time for the light to get to us. He argued that we have no reason for believing this to be the case, so instead the universe can't be bigger than a size that allows for the gaps we see.

We now believe his discarded speculation was right. There is good evidence that the universe is 13.7 billion years old, so we can only ever see light that has been travelling for 13.7 billion years or less. Because light travels at a finite speed, any light that would take more than 13.7 billion years to reach us wouldn't have arrived yet. To fill in all those gaps in the night sky, we would need light to reach the Earth that set off far longer in the past.

Explaining the darkness is entertaining, but on the whole astronomy is all about understanding what is up in the sky rather than what isn't. And once we have the basic building blocks of stars and planets it is worth looking at the most familiar collection of these elements: Our own solar backyard.

chapter 4

Our neighbourhood

To take in and explore our own neighbourhood we need to go back to the relatively small picture of the solar system, the gravitationally linked grouping that consists of our own star, the Sun, and the bodies that orbit it. Galileo might have added some detail like Jupiter's moons, but the seventeenth-century solar system was still pretty much what could be seen with the naked eye: the Earth, the Sun and Moon, five more planets and occasional visitors like comets. As technology has improved we've added more to the picture, and we now understand what's going on a whole lot better.

The Sun, of course, has always been considered special, with good reason. It brings light and heat to the world and was often worshipped as a god. And that reaction came about without anyone realizing just how much the Sun really does for us.

A montage of the planets

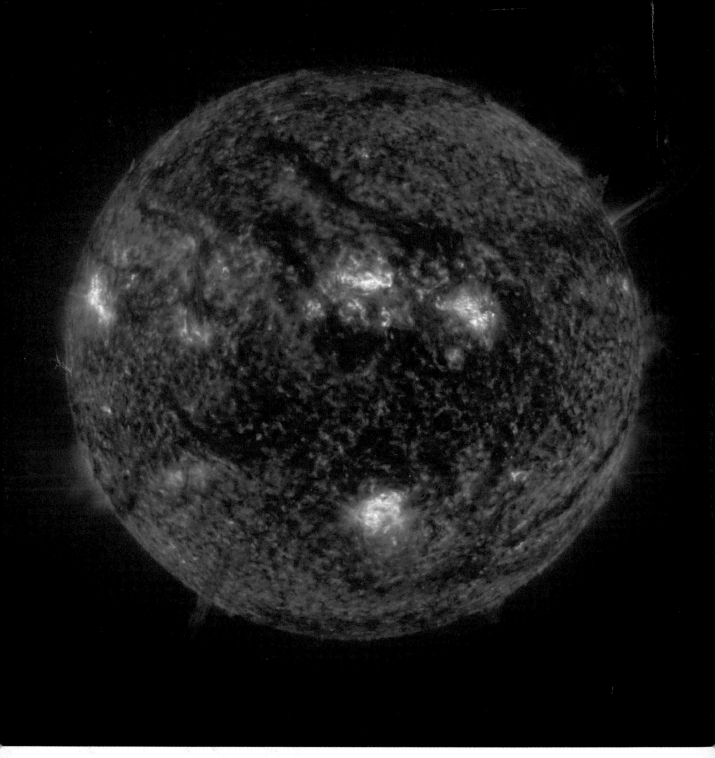

The Sun, an average star, but responsible for the formation of the Earth

Our home star is multiply responsible for the existence life on Earth. As well as giving us light to see by and warmth to keep us alive, the Sun drives the weather – without it, for example, there would be no rainfall – and it is light from the Sun that powers the oxygen-producing process of photosynthesis in plants that gives us the air we breathe. Most fundamentally, without the Sun, the disc of dust from which the Earth coalesced would never have formed. The Sun is, in effect, the creator of the Earth.

Given how wonderful the Sun's bounty is, it can be quite surprising to learn that the Sun is just an ordinary star – impressive, certainly, but a single member of a club with countless others. It's quite a leap to equate this intensely bright, hot beacon in the sky with the tiny, cold points of lights that are stars, but the idea came about surprisingly early. The Italian philosopher friar Giordano Bruno, born nearly twenty years before Galileo in 1548, suggested that every star was another Sun, and even said that the stars could (and almost certainly did) have their own planets like the Earth.

Bruno was charged with blasphemy and heresy, and in the year 1600 was burned at the stake by the Inquisition. While it's true that promoting the idea of many worlds travelling around distant suns was one of the charges brought against Bruno, it would be inaccurate to suggest he was burned because of his views on science. The church was much more concerned with his denial of basic Christian doctrine. But there is no doubt that the cosmological ideas that Bruno wrote about were well ahead of his time.

Gradually, in the fifty years after Bruno's unfortunate death, as the idea that the Sun was at the centre of the solar system gradually took sway, so Bruno's idea that the stars were other suns gradually became accepted. By the time Isaac Newton was applying his ideas on gravity to the universe, it was a natural assumption for him that the stars were all massive bodies like the Sun, rather than tiny points of light in some heavenly sphere.

As a star the Sun is really rather... average, not in a bad way, but it sits around the middle of many aspects of being a star. At around 4.5 billion years old it is about halfway through its active life. It isn't particularly large – technically it is classed as a yellow dwarf – but equally it is not a tiny star. There are hotter and cooler stars out there too.

Oddly, although we naturally make the Sun yellow if we draw it, that 'yellow' designation doesn't refer to the colour the star appears to be. The light from the Sun is white – we just see it as yellow because some of the light spectrum from the Sun, particularly the blue end, is scattered by the Earth's atmosphere (this scattered blue light is why the sky is blue). With the blue end removed, the Sun looks yellow, or red when it's low in the sky and its light passes through more air, which scatters more of the photons.

The reason the Sun is a designated a yellow dwarf is because yellow is the strongest of the colours in the light spectrum it produces, something that isn't obvious while looking at it from space without air to scatter the light (not that you should look at it directly under any circumstances). The Sun's light is so strong that it quickly causes serious damage to the eyes. This is true even when the Sun is partially obscured – hospitals get a steady stream of people with permanently damaged eyes who have looked directly at the Sun during solar eclipses, when the Moon passes in front of the Sun.

The vital statistics of the Sun reflect its importance in the solar system. It may be an average star, but it's anything but an average part of our neighbourhood. The Sun is nearly 1.4 million kilometres across, a diameter over a hundred times bigger than the Earth, and its mass is 330,000 times greater. If you look at the whole of the solar system, over 99.8 percent of its mass is contained within the Sun. And, of course, that star is hot. The outer layers of the Sun are around 5,500°C. If this sounds surprisingly cool, temperatures are thought to get up to around 10 million degrees in the depths where the fusion reactions keep the Sun alive.

Like all stars, the Sun continues to burn using the fusion process described earlier (see page 39). Every second, around 600 million tons of hydrogen ions are fusing into helium. The energy produced by this process is immense, around 400 billion billion megawatts. To put that into context, just 89 billion megawatts of this hits the Earth, yet that is still more than five thousand times our global energy consumption.

Basking in this powerful glow are the Sun's children, the planets. Nestling closest is Mercury. It's the smallest of the planets, but not as tiny as you might think – about a third the diameter of the Earth. Like all the inner planets, Mercury is rocky and suffers from drastic variations in surface temperature. As you might expect for a planet so close to the Sun, temperatures can rise as high as 420°C on the sunlit side.

Mercury lies between a third and a half of the Earth's distance from the Sun – averaging around 50 million kilometres, but this distance varies by as much as 23 million kilometres, more than five times the variation in the Earth's orbit. Although all the planetary orbits are ellipses, the Earth's is nearly circular, while Mercury's is much more squashed. The high temperature reflects both the planet's proximity to the Sun and its slow rotation. Mercury's day lasts around 180 Earth days, so the Sun stays on parts of the surface for long periods of time.

Daytime may be on the warm side, but when night falls, with no blanket of atmosphere to hold the heat in, the temperature plummets, falling as low as -180°C. This is around the temperature at which oxygen becomes liquid. Close up, Mercury looks a bit like the Moon – without an atmosphere to slow down incoming matter or to weather the results of collisions, it is pockmarked with the craters of many meteor strikes.

Because Mercury is so close to the Sun (and so small) it can be difficult to see. It always stays near the Sun in the sky, so it is only visible at dusk or dawn as a faint speck of light. In fact it's quite surprising that early astronomers noticed it – yet it says something for their interest in the skies that Mercury seems to have been recorded in Babylonian tablets dating back over three thousand years, and it was certainly known by the ancient Greeks, although they initially thought it was two separate planets, Apollo, which appeared at dawn, and Hermes, that was seen at dusk.

Mercury shares this morning and evening association with the next planet out, Venus. Because Venus is also closer to the Sun than we are, it never strays too far from the Sun in the sky either. Although it is much brighter than Mercury, Venus can be seen with

ease at the right time of day, even in daylight, and is the brightest object in the night sky after the Moon.

For many years, Venus was considered a sister planet to the Earth. If there were other intelligent life forms out there in the solar system, it was pretty certain they would be on Venus and plenty of science-fiction stories told of the lush vegetation the planet might support – or even little green men. After all, Venus is similar in size to the Earth, with a similar gravitational pull. On the downside, Venus is quite a lot closer to the Sun than we are – about two thirds of Earth's distance – but it has thick clouds to protect it, and it was felt that being on Venus would be like being in a hot tropical forest on Earth – and it could possibly be just as rich in life forms in reality, just as it was in the stories.

It was something of a shock, then, when the first successful probe, Mariner 2, reached Venus in 1962 and started to send back data that uncovered the reality of this hellish planet. The surface temperatures on Venus average around 460°C and can reach as high as 600. It's significantly hotter than Sun-hugging Mercury. The metal lead would run liquid on the surface of Venus.

The planet does have an atmosphere, which was one of the reasons it was thought to be a hopeful home for life, but it is nothing like our atmosphere. Venus is swathed in a thick layer of carbon dioxide, giving it an immense atmospheric pressure at the planet's surface. It is over ninety times the pressure on Earth. The over-the-top greenhouse effect that all the Venusian carbon dioxide produces is responsible for those incredibly high temperatures.

We tend to think of the greenhouse effect as a bad guy because of its impact on global warming. In moderation, though, the green-house effect is highly beneficial. If there were no greenhouse effect on Earth, temperatures would average about thirty degrees lower (around -18°C) and there would very little that could survive except specialist bacteria that thrive in extreme temperatures. But too much of a greenhouse effect certainly is a bad thing, as Venus demonstrates.

We never see any of the features of Venus in the way that you can when looking at the Earth from space. This is because the

Without an atmosphere to protect it, Mercury is pockmarked with meteor strikes

Venusian clouds are so thick, providing a permanent near-white screen to what's below. It's not the CO_2 that we see though because carbon dioxide is transparent. As if to emphasize just how hellish Venus is, those bright white clouds that conceal the surface and make it so bright are that vitriolic substance, sulphuric acid. With such a thick covering, despite being nearer to the Sun than Earth, less light gets through to the planet's surface in the immensely long Venusian day.

It's a strange day because of the way Venus rotates, which is in the opposite direction to the Earth and most of the other planets. The day on Venus, in terms of a single complete rotation, takes 243 Earth days, which is longer than the 225 days it takes to get around the Sun. Because of the way the planet spins, combined with its movement around the Sun, the time between sunrises is just under 117 days.

Like Mercury, the early astronomers assumed that Venus was two separate stars, one when it was seen at nightfall and the other as morning broke. The early Greeks named the morning star Phosphorus and the evening star Hesperus, although by the fifth century BC they had worked out that it was the same planet they were seeing. There was no such confusion, however, over the nature of the next planet from the Sun – the Earth.

It wasn't really until we had space flight, and could see the Earth in all its glory, that we appreciated what a remarkable planet it is. It was no surprise, though, that it was a sphere. We've known our planet is not flat for a long time. The idea that people thought the Earth was flat in medieval times simply isn't true. This commonly held belief seems partly inspired by the strange maps that were produced at the time. These were more philosophical than geographical, putting Jerusalem at the 'centre' of a circular picture of the world.

In practice, any seafaring nation would be aware of the curvature of the Earth. When onboard a ship, it appears as if other ships and distant ports rise up over the horizon, whichever direction you look. What's more, you can see further if you are up on the mast, spotting a ship further around the 'bend' – clear indicators of the

In this radar image of Venus the colours are artificial, indicating altitude
FOLLOWING SPREAD: The colours are simulated in this radar view of the 8 kilometre high Maat Mons on Venus.

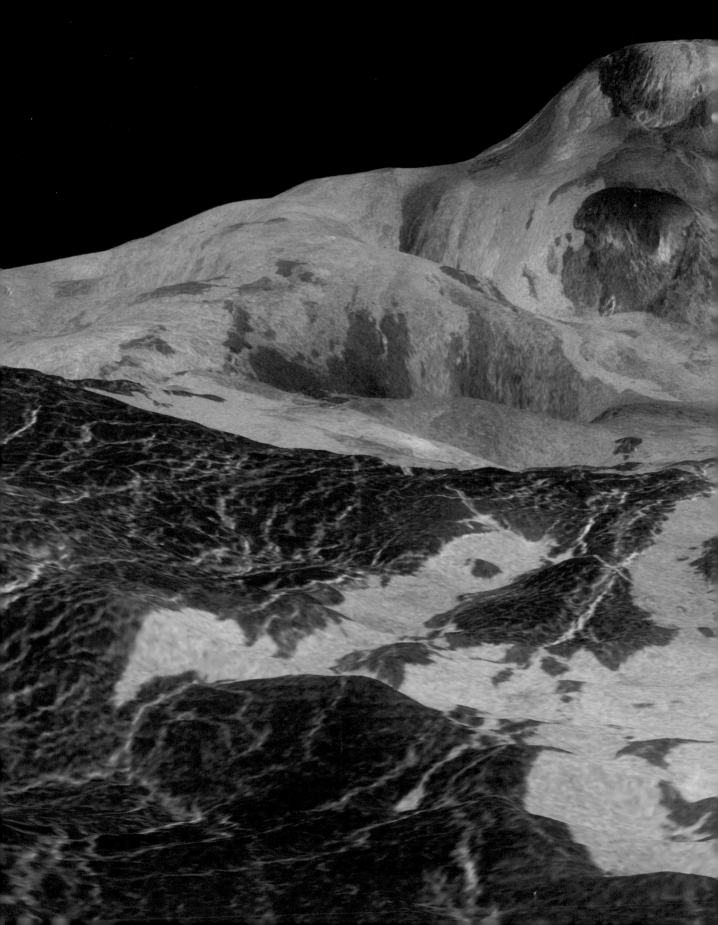

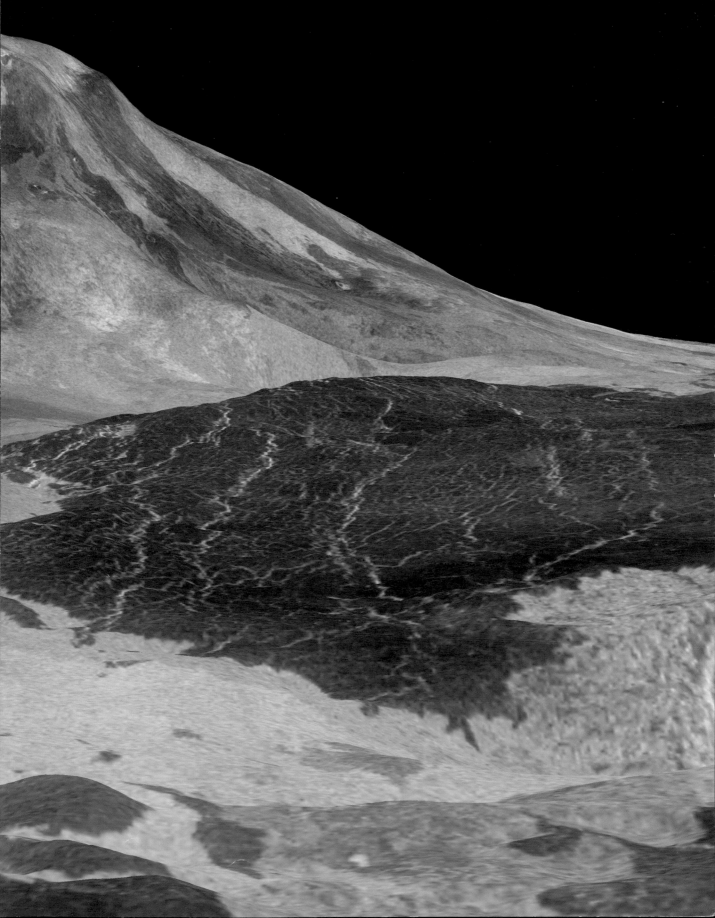

Earth's curvature. The ancient Greeks knew the Earth was spherical (technically it's not, as it bulges in the middle, but the picture isn't bad) and that has been the accepted view among educated people ever since.

However, it is only when we were able to take photographs from space that it became possible to see the way the Earth differs so drastically from the other rocky planets with the striking blue of the seas and the green vegetation on the land masses. It is obvious at first sight that this planet, sitting in what is described as the 'Goldilocks Zone' where temperature, solar radiation and much more are just right, is something special.

We don't need to say too much about the Earth itself – it's kind of familiar – but the Moon is of particular interest. The Earth is the first of the planets working out from the Sun to have a moon, and in our case it's no ordinary moon. It is massive. Although it isn't the biggest moon in the solar system – it comes in at number five, with the number one being bigger than Mercury – with one arguable exception (which we'll come to later) it is the biggest moon in comparison with the planet it orbits.

The Moon is 3470 kilometres in diameter, around a quarter of the width of the Earth. When we see it at night, like the planets, our sole moon is lit by reflected light. The Moon's surface is mostly dark grey, but the light from the Sun is so powerful, and the night sky so dark by comparison, that it appears a milky white colour. Although moonlight is bright enough to read by, this brightness is partly an illusion provided by the flexibility of our eyes – moonlight is about 300,000 times weaker than sunlight.

When we look up at the Moon we see a familiar change of shape as it goes through its waxing and waning. The phases of the Moon as it moves from the practically invisible new Moon, through crescent to full and back again simply reflect the angle at which the Sun's rays hit it. Both Venus and Mercury also have phases; in the case of Venus this can easily be seen with a pair of binoculars. But however much of the Moon is illuminated, we always see the same face. The so-called dark side is never on show.

The far side of the Moon is of course no darker than the side

Our moon is the largest in the solar system in comparison to the planet it orbits

we see. When there's a new Moon, so the side facing us is in darkness, the far side is fully illuminated. But we never get to see that side of the Moon except from space probes because the Moon happens to rotate at just the right speed to keep the same face pointing towards the Earth. This wasn't always the case, but tidal forces (more about these in a moment) have dragged it into synchronicity.

This rotational coincidence has a reason, then, but there is another coincidence about the Moon that is entirely down to chance – it looks almost exactly the same size in the sky as the Sun. This match up varies a little, depending on the distance between the Earth and the Sun, but roughly speaking the Sun is four hundred times bigger than the Moon and is also four hundred times further away. The result is that when the Moon comes between the Earth and the Sun, it completely covers the face of the Sun and we get a total solar eclipse.

These don't happen every time there is a new Moon (with the Sun illuminating the far side of our satellite) because the Moon's

orbit around the Earth isn't in the same plane as the Earth's orbit around the Sun. It is only when the three bodies are positioned just right that a solar eclipse occurs. Lunar eclipses, when the Earth comes between the Sun and the Moon, are much more common, and are visible across the night side of the planet. In a lunar eclipse the Moon doesn't go entirely dark – this is because the Earth's atmosphere sends some of the sunlight bending in towards the Moon, turning the eclipsed Moon blood red.

The size that the Moon appears to be in the sky is highly misleading. Our brains seem to assume the Moon is closer than it really is, particularly when it is low and near to objects like buildings or trees – we have all seen and wondered at a particularly large Moon. In reality, with all optical illusions accounted for and removed, the Moon's visible size is not even as big as a penny held at arm's length. It is more like the size of the hole produced by a hole punch, held at that distance.

While we can't be a hundred percent certain how the Moon came into being, there is evidence to show that the Moon was formed around 4.5 billion years ago, perhaps 50 million years after the solar system as a whole first formed. It is thought that a stray body the size of Mars collided with the still-forming Earth, blasting off a huge chunk of material that partly came from the Earth and partly from the colliding body.

The Moon isn't just a pretty face – it is also responsible for the tides we experience (the Sun, too, makes a contribution to the seasonal variation, but the Moon is the main culprit). As the Moon orbits the Earth, each has a gravitational impact on the other. The Moon pulls the Earth and everything on it towards itself. The side of the Earth facing the Moon is the closest, so the water on that side bulges out towards the Moon.

There are, however, two tides a day. Some of the power for the tides comes from the way the Earth and the Moon dance around a point between them, rather than the Moon purely orbiting the Earth. This oscillating movement of the Earth, rather like a theme park ride, has the effect of pushing out a tide on the side facing away from the Moon.

Solar eclipses are dramatic because of the coincidence in apparent size of the Sun and Moon.

If the Moon had water on its surface the tides there would be vast. The force of gravity increases with the mass of the body, and the Earth is eighty times heavier than the Moon, so it produces eighty times the gravitational tidal effect. This might seem too big, as the gravity on the Moon's surface is about one sixth of that on the Earth – as demonstrated by those bouncing lunar astronauts. However, there is another component coming into play.

The pull of gravity you feel involves the mass producing the pull divided by the square of the distance you are from the centre of that mass. When comparing tidal forces we are dealing with the same distance between Earth and Moon. But on the surface of the Moon you are 3.6 times closer to its centre than you are to the centre of the Earth when standing on our home planet's surface. This means there is a stronger force than the mass of the Moon predicts, producing a gravitational pull that is one sixth of that on the Earth.

Travelling still further from the Sun, we reach Mars. Although nowhere near as bright as Venus, or even Jupiter, Mars is quite distinctive when seen in the sky because even with the naked eye it has a clear red colour that makes it stand out from the surrounding

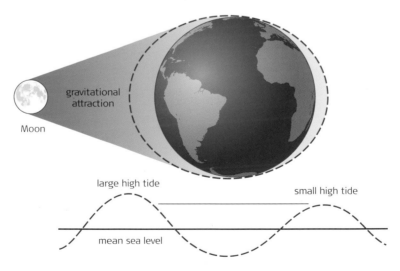

The Moon is the primary influence on the Earth's tides

stars. The last of the rocky planets, Mars is the nearest thing we have to another hospitable planet in the solar system (moons are a different matter, as we'll see later). Sadly, this isn't saying much.

On average Mars is around 70 million kilometres further from the Sun than the Earth (even at it's closest it's more than 50 million kilometres), which is why the idea of a manned mission to Mars is so much more unlikely than another mission to the Moon. The distance from Earth to Mars is around a hundred and forty times further than the distance to the Moon.

Mars is about half the diameter of the Earth and has a tiny fraction of the atmospheric pressure. Surface temperatures, however, are a lot more hospitable than is the case on the inner planets. Weather conditions on Mars vary from an admittedly chilly -90°C to around 20°C, spending most of the time below freezing. But this temperature range does not make the survival of bacterial life impossible, especially if that life was present under the surface, a possibility reinforced by some evidence for the presence of water on Mars.

Accompanying the planet named for the Roman god of war are two moons with scary names: Phobos (fear) and Deimos (dread). These are the first bodies we've come across in our trip through the solar system that aren't even vaguely spherical. They look more like random lumps of rock and may well be captured asteroids. Compared with our moon they are both tiny – Phobos is about 26 kilometres long on its longest axis, while Deimos is around 15 kilometres – not even as big as a good-sized island.

Celsius

500°
400°
300°
200°
100°
0°
-100°
-200°

Venus
Mercury

Earth

Mars

Jupiter

Saturn

Neptune

Uranus

Planets not shown to scale

Average temperatures across the solar system

The Hubble's dramatic shot of Mars shows markings once thought to be canals

When nineteenth-century astronomers trained their telescopes on Mars they thought that they could see a series of channels, soon referred to as 'canals' in a mistranslation from the Italian. This led to speculation that an intelligent civilization on Mars had built these canals thousands of years ago and that perhaps now sheltered underground because their inhospitable planet had lost most of its atmosphere. It was from this misunderstanding that the idea of Martians as the typical alien invader came about, reinforced by H. G. Wells's striking book *The War of the Worlds*.

The canals simply don't exist. There are one or two long canyons on Mars, but nothing that could be interpreted as canals. This was before photographic technology was used in astronomy. The observers of the time would look through a telescope and then try to transcribe what they saw in the form of a sketch. It still seems remarkable that more than one astronomer could imagine the same structures, but once one had drawn them, they seem to have had a strong psychological influence on others.

In recent years Mars has been studied both by orbiting probes and by unmanned landers. There is no evidence of life yet, but there is certainly an interesting landscape. Mars has huge volcanic mountains, including Olympus Mons, which, at around 22 kilometres in height, towers nearly three times higher than Mount Everest. There are also vast canyons, like Valles Marineris, which, at 4,000 kilometres long and around seven kilometres deep, doesn't beat Earth's biggest underwater canyons, but makes the Grand Canyon seem like a scratch.

For many years now, Mars has been the focus of efforts on a manned mission to another part of the solar system. It presents significant challenges, however. Because it is a hundred and forty times the distance to the Moon, a journey there would take months, not days. This means astronauts would be exposed to considerable levels of radiation, would be weightless for a length of time that would cause considerable damage to their muscles and bones, and

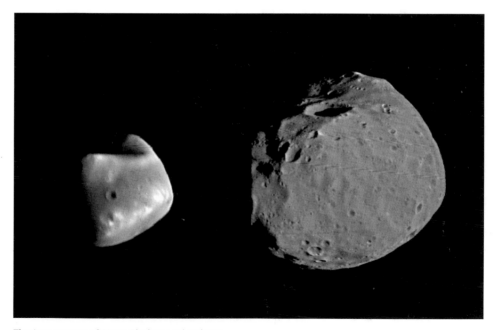

The two moons of Mars, Phobos and Deimos

The vast volcanic Olympus Mons on Mars is 22 kilometres high

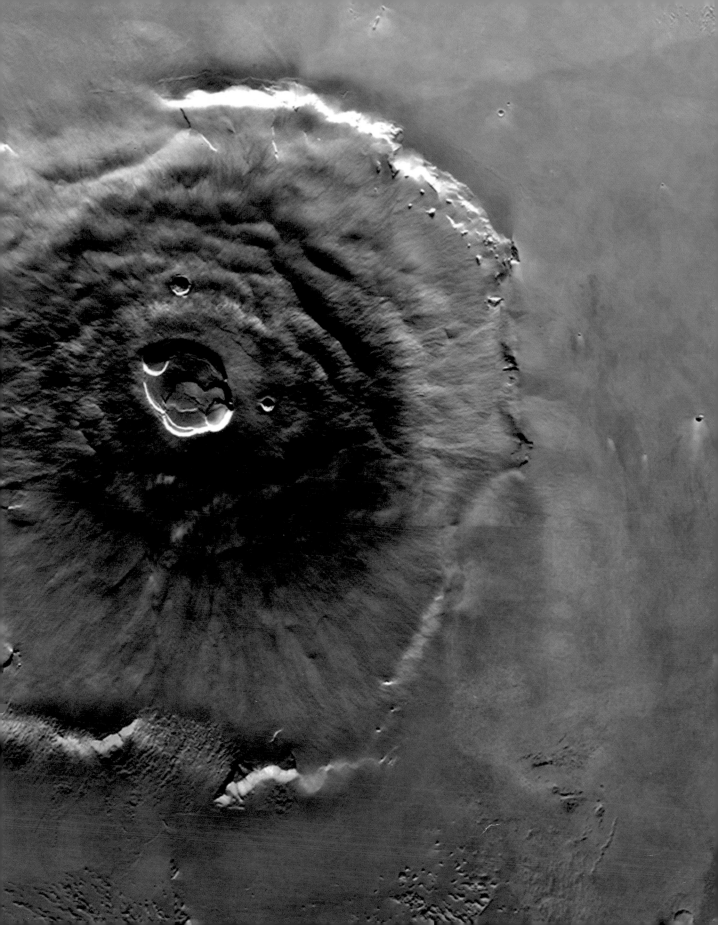

would have to cope with being cooped up in a confined space for long periods.

This doesn't stop our efforts towards a manned mission, though. In 2011, a Russian 'expedition' underwent a simulated trip to Mars by being sealed in a capsule for the appropriate period of time, while NASA has had the long-term goal of a manned mission to Mars since 2004, with the aim of making the journey in the late 2030s, a similar timescale to a potential ESA mission.

Such an expedition could not be a colonizing trip. With its negligible atmosphere, Mars is not yet a viable long-term habitation for human beings. It is true that a manned mission to Mars would make it easier to undertake some experiments, but many scientists believe that a manned mission is both too risky and would produce nowhere near as much useful scientific data as sending more unmanned probes. As these can be sent for a fraction of the price (and risk), we could launch hundreds of unmanned probes if we don't pay for a manned programme.

With such a thin atmosphere Mars remains a raw wilderness, albeit one that can present us with beautiful images. Further out from the Sun we meet more ugly lumps of rock not unlike the moons of Mars. This asteroid belt was once thought to be the wreckage of a planet that had exploded, a lost planet that had the same romantic attraction as the mythical continent of Atlantis. But those who believed in an exploding planet had got the picture back to front.

What seems to have happened is that as the disc that spun around the young Sun gradually coalesced into planets, the area between Mars and Jupiter was so disrupted by the gravitational forces from Jupiter that there was no possibility of a full-sized planet forming and the result, instead, was a broad band of rubble, varying in size from dust particles to the biggest asteroids, chief of which, Ceres, is 950 kilometres across – many times bigger than the moons of Mars.

Although a fair amount of the material from the belt has been lost – some of it by streaking across the solar system as meteors – what remains is nowhere near enough to form a good-sized planet. Although there are vast numbers of small asteroids in the belt, they don't form the kind of nightmare maze that appear in science-fiction movies and video games, where, without expert piloting, passing through inevitably leads to a collision. As yet none of the nine probes that have gone past Mars have run into trouble in the belt. There is just an awful lot of space between solid objects.

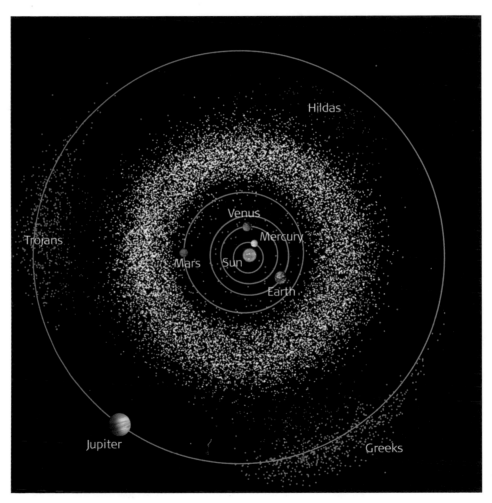

The asteroid belt between Mars and Jupiter is no longer thought to be an exploded planet

Meteors, incidentally, are what we see as shooting stars on Earth. They range in size from dust particles to good-sized boulders. Technically, the lump of substance while in space is called a meteoroid – meteor is the visible trail in the sky. Most meteors burn up due to friction when they enter the atmosphere but the bigger ones will land, where they become known as meteorites. All the planets and moons are subject to meteor collisions – some, like the Moon, show many obvious craters. These are less obvious on the Earth where few get through the atmosphere and those that do are eroded by the weather and earth movement.

Back in the asteroid belt, just how much space is out there can be seen by looking at the scale of the region. It stretches between about 330 and 480 million kilometres from the Sun – a similar depth to the distance from the Earth to the Sun – but occupies a much larger volume than the space within the Earth's orbit because it starts so much further out.

And beyond the asteroid belt? Next we reach the bruiser that prevented another planet forming, the biggest planet in our solar system, Jupiter. This is a different kind of planet from anything we've met so far – in fact, you could almost look at Jupiter as an object that is halfway between a planet and a star. Like most stars it is primarily composed of gas, and it is vastly bigger than a planet like Earth, although it's too small to have enough temperature and pressure to slip over the limit and ignite into fusion.

The numbers are one thing – Jupiter is over 140,000 kilometres in diameter and weighs in at over 1.8 billion billion billion tonnes – but what puts it into perspective is its most familiar feature, the great red spot, a vast storm that seems to have been raging for at least four hundred years. It is so large that you could fit the Earth into it a couple of times over.

It is thought that Jupiter may have a relatively small core of rock buried deep inside, but most of the planet is made up of the elements hydrogen and helium – hence the term 'gas giant', which is applied to Jupiter and its three fellow outer planets. A fair amount of the hydrogen in the inner part of the planet will be in liquid form. This is a special kind of material, so-called metallic hydrogen. It's

The great red spot on Jupiter is a vast storm, twice the diameter of Earth

A composite of Jupiter's four largest moons (Jupiter is not to scale)

still a liquid (as any metal can be at the right temperature), but like a metal its electrons are relatively free to move around, conducting heat and electricity.

However, just because Jupiter is mostly liquids and gasses doesn't make it a bland, featureless ball. Apart from the great spot, the planet is ringed with different coloured bands and swirls, clouds that are mostly made up of frozen crystals of the nitrogen and hydrogen compound, ammonia. The atmosphere has several more elements present – if there was some way to visit and float on the 'surface' of Jupiter, among the vast lightning bolts, hundreds of times more powerful than we experience on Earth, you might see glowing pink neon rain.

Another way of appreciating the size of Jupiter is that if you lumped together all the other planets you wouldn't reach half the mass of this monster. (And bear in mind that with so much hydrogen, Jupiter isn't particularly dense.) Even though Jupiter is over 750 million kilometres from then Sun, making it around 600 million kilometres away from us at the closest point in the two planets' orbits, it is often the second brightest object in the night sky after the Moon when Venus is below the horizon. Mars can top it at its closest, but Jupiter appears a clear white colour to the naked eye and is easily distinguished.

Jupiter was also the first planet, apart from Earth, that was known to have moons. When Galileo trained his early telescope on Jupiter, as we have seen, he discovered four bodies orbiting it. Now we know that there are many more. Jupiter's massive gravitational pull is very good at capturing stray rocks. There are 63 moons that have been given names (it's hard to give an exact number when you get down to small captured asteroids). The big four moons, those spotted by Galileo, are, in decreasing order of size, Ganymede, Callisto, Io and Europa.

To put them into scale, all but Europa are bigger than are own Moon, which is unusually large in its own right. The biggest, Ganymede, is over 5,200 kilometres across (compared with the Moon's 3,400 kilometres), making it bigger than the planet Mercury. You might expect that these natural satellites would be boring

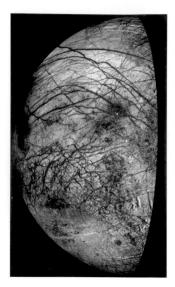
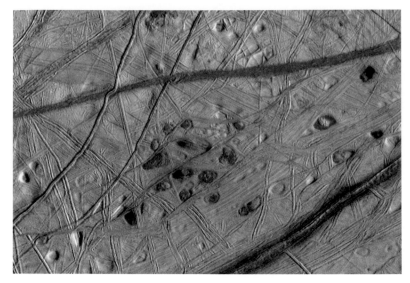

Europa's scarred icy surface covers a liquid ocean that makes this moon of Jupiter a likely candidate for extraterrestrial life

lumps of rock, but there is considerable hope that at least one of the moons of Jupiter could harbour bacterial life.

All but Io of the big four are thought to have oceans of water well beneath the surface. Despite Jupiter being so far from the Sun, these lunar oceans are likely to be kept warm enough to be liquid by the tidal energy from the planet. The most likely candidate for hosting life is Europa, whose unusually smooth surface is an ice layer which almost certainly covers a liquid ocean, while any water on Ganymede and Callisto would have to be much deeper under a rocky surface.

Jupiter's next neighbour out from the Sun is the pinup of the solar system, Saturn. It's the last of the planets that can be made out with the naked eye. It has been known since ancient times, and is the second biggest after Jupiter at around 100,000 kilometres in diameter. The obvious characteristic that makes Saturn look so remarkable is its rings. Around the same time that Galileo discovered the four large moons of Jupiter, he also trained his telescope on Saturn and saw something strange.

Galileo couldn't work out what to make of the strange misshapen appearance that Saturn took through the telescope, sketching it with what appear to be cartoon ears – instead, as we now know, Saturn has its huge and magnificent rings. It's not the only planet to have them – Jupiter and Neptune both have faint rings and Uranus has a more noticeable set – but no planet in our solar system compares with Saturn in magnificence.

Not surprisingly, like Jupiter, Saturn has a good collection of moons – 53 of them have names – with some comparable in size to Jupiter's. The biggest, aptly named Titan, is only a little smaller than Ganymede and beats Mercury with a diameter of around 5,000 kilometres. That's 40 percent the size of the Earth. Titan is particularly interesting because it is the only moon that we know of with a thick atmosphere, which is mostly nitrogen at a higher pressure than Earth's. But even the greenhouse effect from the methane present in that atmosphere only lifts the surface temperature to around -180°C.

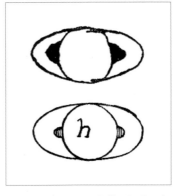

Galileo's sketches of Saturn

Saturn itself, like Jupiter, is mostly hydrogen and helium, with a higher percentage of hydrogen, and a coating of ammonia clouds that give it a banded appearance. The rings are made up of lumps of water ice with carbon impurities. It's not certain how they formed – they could be material from the original disc that formed the solar system, which never had a chance to come together to form a moon as it spun around the newly formed planet, or they could be the result of a moon breaking up. Whatever the cause, they are stunning in appearance, especially as seen from the probes that have travelled past Saturn.

Apart from the stars, Saturn completed the main contents of the universe in the view that held all the way through to the eighteenth century. It was then that a musician and amateur astronomer, William Herschel, would add another planet to the catalogue. He initially called this George's star (Georgium sidus), in an effort to ingratiate himself with the English king, but the new planet was to become known as Uranus. (Herschel still did pretty well, being given the paid appointment of King's Astronomer.)

Herschel saw Uranus as a tiny star that moved across the

Saturn's rings in all their glory

Closeup of Saturn's rings showing the shadow of the moon Mimas

heavens. It's not surprising as, despite being around 50,000 kilometres across (about half the size of Saturn and four times the diameter of the Earth) it is typically over 2,800 million kilometres from the Sun, making it twenty times further from the centre than the Earth is. When probes first took close-up photographs of Uranus they provided a picture that looks very different from the first two gas giants. Here there is none of the striking banding in a planet that is a uniform light blue in colour. Although still based mostly on hydrogen and helium, more compounds like water, methane and ammonia make up Uranus.

Seen in a photograph, Uranus looks ice cold – and that's not inappropriate. With temperatures falling as low as -220°C it is, as far as we know, the coldest of the planets. This far from the Sun, most of the heat that keeps a planet above the -270°C of space comes from the planet itself, but Uranus seems to have very little residual heat left over from its formation. It is joined on its cold journey by at least 27 moons, two of which, Titania and Oberon, were spotted by Herschel. Titania is the biggest, and at 1500 kilometres across is about half the size of our Moon.

Uranus is slightly bigger than the last of the planets, Neptune — but the existence of Neptune was not confirmed until 1846. At 4,500 million kilometres from the Sun, thirty times as far as the Earth, Neptune is in the far reaches of the solar system. Like Uranus, Neptune is blue and has a composition that includes a more diverse mix of compounds than its neighbour, giving the planet more surface banding effects.

Neptune has at least 13 moons, but only one that is large enough to not look like a misshapen piece of rock. This largest moon is Triton, discovered within weeks of Neptune itself. At 2,700 kilometres in diameter it is a little smaller than our Moon, but it is still sizeable, and, unlike any of the large moons around the other

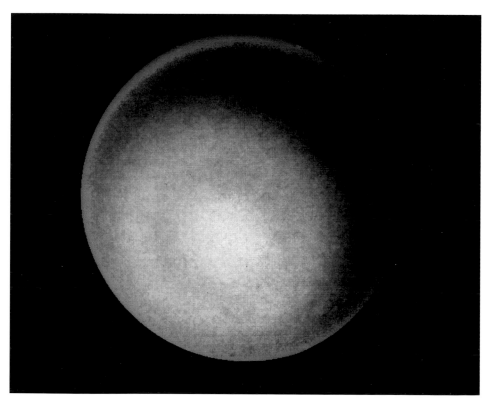

The icy blue Uranus, the coldest planet

planets, it orbits Neptune in the opposite direction to the planet's rotation.

Before 2006, we would have gone on to add a ninth planet, Pluto. Discovered in 1930, Pluto really doesn't fit well with the pattern of the solar system. It is a rock and ice structure about 2,300 kilometres across (significantly smaller than the Moon) with a wildly varying orbit that takes it between 4,500 million and 7,400 million kilometres from the Sun. Pluto is said to have a sizable moon, Charon, but it's more accurate to say that they are a pair of minor planets orbiting each other (a 'binary system').

Although some were upset when Pluto lost its status as a planet, it makes a lot of sense to demote it. It fits better in the category 'minor planet' like the large asteroid Ceres and the recently discovered minor planet Eris, which is bigger than Pluto. Once we get beyond the orbit of Neptune there are broadly three bands of object that are still under the sway of the Sun, although they are decreasingly influenced by our star as they are so distant.

First comes the Kuiper Belt. In effect this is a second, much bigger asteroid belt that extends out from Neptune for around 3,500 million kilometres. There are over a thousand clearly identified objects in the belt (no doubt there are millions more), ranging in size from Eris to fragments of ice. Although some of the Kuiper Belt's inhabitants, like Pluto, are rocky, many are frozen compounds, similar to those in the atmosphere of Uranus and Neptune – water ice, methane and ammonia.

Beyond the Kuiper Belt we reach the scattered disc. This is a much less populous region, more unstable than the Kuiper Belt at its greater distance from the Sun's gravitational influence. The scattered disc objects are all icy, with some of them in such extreme orbits that they come plunging in towards the Sun to form comets.

Comets have been familiar objects in our skies for thousands of years. The name comes from the Greek for 'wearing long hair' because comets typically have a tail, a fuzzy band that stretches away from the Sun. Comets are made up of ice and dust, and the blast of particles that streams from the Sun called the solar wind abrades their surface, leaving the trail behind. Comets typically

Neptune's blue sphere features more detail than Uranus
FOLLOWING PAGES *Size comparison of minor planets Eris and Pluto, and the asteroid Ceres*

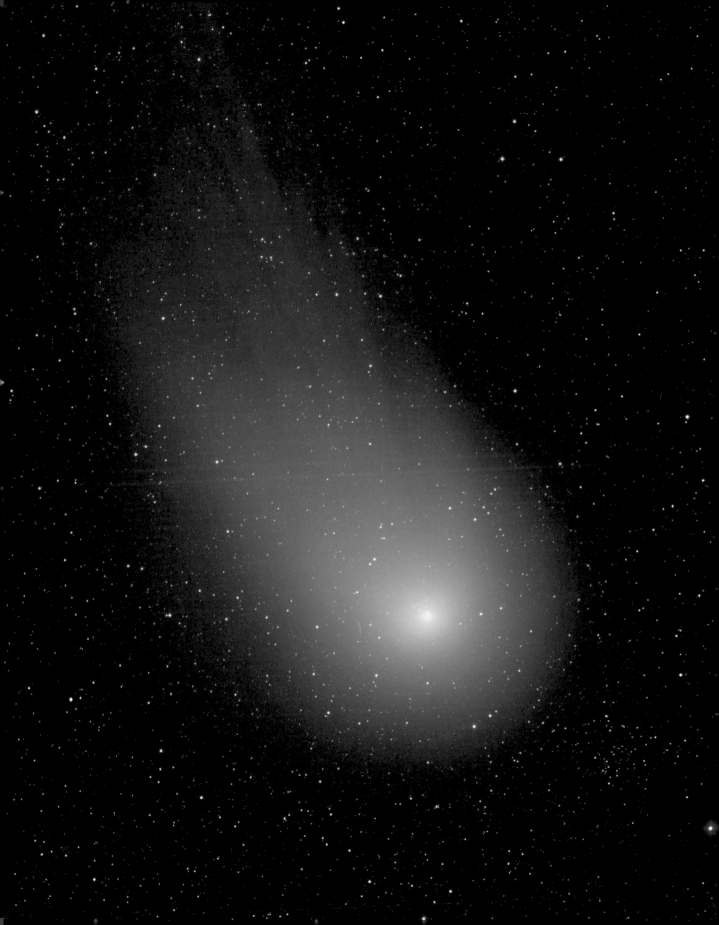

*Comet Wild 2
photographed by
NASA's Stardust
spacecraft*

have extreme elliptical orbits, travelling from the outer regions of the solar system to pass around the Sun closer than the Earth, then departing again for the outer limits.

Finally in our survey of the solar system, at the extremes of the Sun's influence, there is (probably) the Oort Cloud. Here we are so far from the Sun that we can be 50,000 times further away than the Earth. Sunlight takes up to a year to reach this hypothetical cloud of icy objects, each a potential comet. The Oort Cloud's existence is deduced rather than directly observed, but it is thought that the comets with very long periods between appearances originated here, as did some short-period comets like Halley's Comet, which were subsequently thrown off course by the gas giants to adopt a shorter orbit.

By the time we reach the Oort Cloud, the Sun is just another star, three times closer than any other, but still nothing more than a bright spot in the blackness of the universe. From here much bigger vistas open out that make the whole solar system seem incomparably tiny. After exploring our neighbourhood, it's time to go large.

A comet detected by the Near Earth Asteroid Tracking project (NEAT)

chapter 5

A growing space

The solar system is a big place. Travelling at the maximum speeds that the Apollo missions achieved, which were a fairly blistering 40,000 kilometres per hour, it would take nearly 13 years to reach Neptune on the outskirts of the solar system. But to get to the nearest star (other than the Sun), Proxima Centauri, would take over 126,000 years. As for the nearest galaxy beyond the Milky Way, we're talking 1500 trillion years. Better take a packed lunch.

Once telescopes had been developed, so many more stars could be seen that it was hard to continue thinking of the solar system as the limits of everything, with the stars just plastered on a sphere at the edge of the universe. As it became accepted that the vast array of stars in the sky were each suns in their own right, the

The star factory that is the Orion nebula

scale of the universe had to be much bigger than was first thought. Given the way the solar system spreads out through space, why should the stars not also be distributed out at different distances, more massive than anything nearby?

Getting any realistic picture of the scale of the universe proved difficult. It's true that some stars were brighter than others, and a naïve approach would be to say the brighter the star is, the closer it is. We know the Sun is a lot brighter than the other stars, so this sort of makes sense. But brightness could never be used as a simple measure of distance. The stars might all be the same distance away, but differing in brightness – or they could be scattered across a wide range of distances and burning at different intensities. And there were special features in the sky that suggested a more complex structure than just a pattern of scattered stars.

These structures are not constellations. Constellations are handy to be able to recognize which star is which. Orion, for example, is easily spotted because of its distinctive three star 'belt' – and doubly useful because it is visible pretty well anywhere in the world as it sits near the equator in the sky. If you want to spot two famous stars – the bright star Rigel and the very red Betelgeuse, they are both in Orion. Rigel is at the bottom right of the four main stars outside the belt, while Betelgeuse is at the top left.

Because the stars aren't always in the same place in the sky, with the rotation of the Earth and its movement around the Sun, it is very useful to have recognizable shapes like constellations. However, that's all they are good for. Constellations aren't groupings of stars, just apparent shapes that our pattern-loving brains make up. Although we think of the stars in a constellation as being 'together' some are likely to be much further from each other than they are from our Sun.

However, there are real groupings and collections of stars, the most dramatic of which, to the naked eye, is the Milky Way. Without a telescope, provided the night is dark enough (sadly, this is difficult these days with the glare of street lighting), the Milky Way is a dusty sprinkling of light that forms a band across the sky. Seen through a telescope it is a vast collection of stars much more tightly packed together than they are in other parts of the sky.

Then there are nebulae. A nebula is a small fuzzy patch in the sky, rather like a star that has been trodden on and squished. Where the Milky Way was named for its likeness to a spill of milk in the sky, a nebula is a fog or mist. For a long time nebulae were thought to be stars in the process of forming (and we now know that some of them are) – but back in the eighteenth century, William Herschel had different ideas.

He suspected that the Milky Way was a vast collection of stars, roughly disc shaped, of which our solar system was a tiny part, just one star of that massive cluster. This picture of the Milky Way gradually became accepted as the nature of the universe – it was something much bigger than the solar system, containing millions of stars, and we saw most of it as the Milky Way. But Herschel had an even more remarkable idea.

Some of the nebulae seemed to have too much structure to be a fuzz of gas that would eventually form a star. He thought he could make out stars in them, and suggested that they were other Milky

The Andromeda galaxy, one of the nebulae Herschel believed were collections of stars

The Milky Way seen alongside the European Very Large Telescope Array at Cerro Paranal in Chile

Ways – bodies that would eventually become known as galaxies (a term simply taken from the Latin form of Milky Way: galaxias). For Herschel, the universe was much bigger than anyone had imagined. But with this idea he was ahead of his time. Even Herschel would come to doubt that such a dramatic possibility could be true, so he switched back to accepting the Milky Way was the universe. But

how big was that universe? How far away were the different stars? Without being able to go out with a tape measure, how was it possible to find out?

The earliest practical way to measure distances to the stars can be determined with a simple experiment. Hold a finger up in front of your face and alternately close your left and right eyes. As you

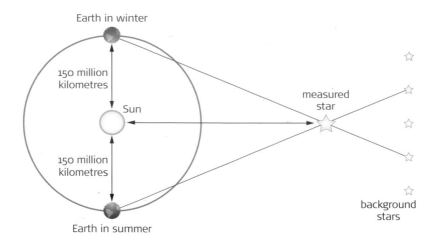

Earth in winter

150 million
kilometres

Sun

150 million
kilometres

Earth in summer

measured
star

background
stars

Observations taken six months apart enable a parallax measurement based on a separation of 300 million kilometres

do so, the finger seems to move against the background. Now hold your finger at arms length and repeat the experiment. It still moves, but not as far. The further away something is, the less it seems to move when you switch viewpoint between your eyes.

You should be able to use the same technique, called parallax, to tell how far away a star is – but if you look at the stars and switch eyes there is no visible shift. It's not surprising, given how far away they are. But imagine your eyes were 300 million kilometres apart. Then the shift would be much bigger. If you look at something in the sky, then wait six months and look again, the Earth will be on the opposite side of its orbit around the Sun, shifting about 300 million kilometres sideways. And this is enough to measure the distance to the nearer stars.

Astronomers even make use of this effect in a unit of measurement. In fact astronomy is by far the least scientific of the sciences when it comes to measurements. All the other sciences have standard units – all distances, for example, are measured in metres. Sadly the distance to even the nearest star is quite messy

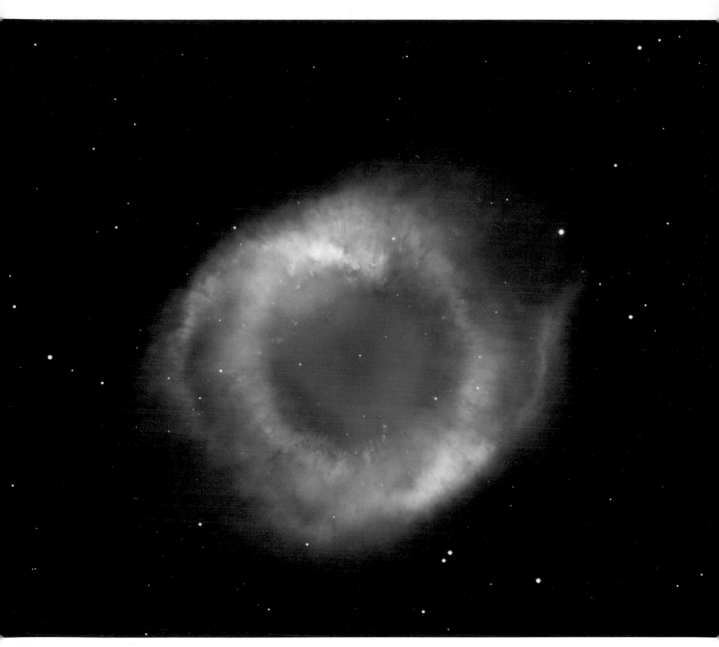

The Helix nebula, glowing gas around a dying star not unlike the Sun in metres. Proxima Centauri is around 39,700,000,000,000,000 metres away. But scientists deal with much bigger values than this by using a simple trick.

Big numbers are represented by the number of zeroes that come after them (technically the number of powers of ten). So instead of using 39,700,000,000,000,000 metres, a scientist would write 3.97×10^{16} – where 10^{16} means 1 with 16 zeroes after it. (If they wanted to show a very small number like 0.00000000009, they would write 9×10^{-11}, meaning 9 divided by 10^{11}.) Astronomers have never got the hang of this. They tend to talk to common folk in light years and to each other in yet another unit.

A light year is the distance light travels in one year. As light moves at 300,000 kilometres per second in a vacuum, this makes a light year around 9,467,000,000,000 kilometres. This is useful, because just a few light years can represent a good-sized distance in space, and the value instantly tells us how far back in time we are looking. When, for example, we look at the Andromeda galaxy, at around 2.5 million light years distant, we are seeing it as it was 2.5 million years ago, long before human beings existed.

For ease of measurement, though, astronomers confuse everyone else by using a third unit, the parsec. This comes back to our trick with the moving finger – parsec is short for 'parallax arc second.' A second of arc is an angle. A full circle is divided into 360 degrees, and these are degrees of arc, of curvature. So one degree is 1/360 of a complete circle. A minute of arc is 1/60 of a degree, and a second of arc is 1/60 of a minute. If an object is one parsec away, then the movement seen in it when the Earth goes from one side of its orbit to the other is one second of arc.

By doing a little geometry, and given the diameter of the Earth's orbit, you can work out that the distance we are from a star that moves by one second of arc is around 3.1×10^{16} metres, which makes it about 3.25 light years. So, uniquely in the sciences, astronomers can confuse you by having not one, not two, but three incompatible units of measurement.

After the initial excitement of measuring the distance of our near neighbours using parallax, it soon became apparent that the

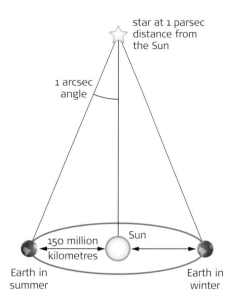

A parsec measures the angle an object appears to shift by as the Earth moves around its orbit

star at 1 parsec distance from the Sun

1 arcsec angle

150 million kilometres

Sun

Earth in summer

Earth in winter

universe was much bigger than anyone expected. There are plenty of stars that don't give any shift at all, even when viewed on opposite sides of the Earth's orbit because they are too far away. Measuring their distances involves a little more guesswork, which meant it took a couple of hundred years before some of the distances were worked out with any accuracy. This was using standard candles.

The idea is simple. If I have two identical candles and position one further away than the other, the more distant candle will look dimmer than the nearer one. If I measure the relative brightness of the two and know the distance to the nearer candle, I should be able to work out the distance to the more distant one. Similarly, if there are two stars in the sky that have the same actual brightness, but one is further away, we can measure the closer one by parallax, and use the difference in brightness to work out how far away the dimmer one is.

That's all very well, but how do we know that the two stars are actually the same brightness? Maybe the dimmer one is the same distance away, but... dimmer. Or a brighter star could be further away than an equivalent star, but brighter still than it appears. Going back to Orion, the brightest star Rigel is three times as far away as the dimmest of the main stars, Mintaka, which is the right hand one of the belt. (Confusingly Mintaka is actually four stars in a complex system.)

Using standard candles meant identifying particular types of star that usually, perhaps always, have the same brightness. Luckily there are distinct types of star, families where brightness can be predicted with some accuracy. Different types of stars have varying combinations of material in them, which can be identified using a spectroscope, a device that looks at the particular colours of light given off by different elements. And other stars are even more distinctive.

The stars that became the first standard candles are variable stars. These are stars that get brighter and dimmer over time, following a regular pattern. The earliest variable stars used as standard candles, called Cepheid variables after the constellation

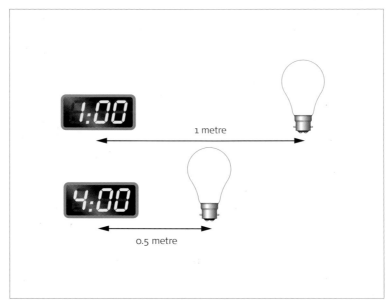

1 metre

0.5 metre

Standard 'candles' – an object becomes fainter with the square of its distance

Cepheus, seem to be in a cycle where sometimes they are puffing up with pressure from the reaction in the star, and sometimes they are collapsing under the force of gravity. After observing many of these stars, it appeared that the speed of their variation was a good indication of their brightness. So if you find two variables stars with the same rate of flashing, but one is always brighter than the other, the brighter one is closer.

By the 1920s, the positions of a great many stars had been measured – but one thing remained uncertain. Was the Milky Way the entire universe – was every star in the Milky Way or floating around its environs? – or was this vast collection of stars just one of many 'island universes' – galaxies spread across a much bigger universe? No star had yet been discovered outside the Milky Way's span, which by then had been established at around 100,000 light years from side to side.

In 1923, an American astronomer called Edwin Hubble came up with the deciding evidence. He was studying one of the best-known nebulae, one that we have already met, the Andromeda nebula. Under the powerful telescopes available by then, the nebula was revealed to be a vast collection of stars. And Hubble spotted the right kind of variable star – a Cepheid variable – to be able to pin down how far away the nebula was. He calculated that it was 900,000 light years distant – far outside the Milky Way. What we now call the Andromeda galaxy, the Milky Way's nearest large neighbour, is every bit as big as the Milky Way itself.

In fact Hubble made a mistake in his distance measurement. There are two types of similar Cepheid variable star, with significantly different brightness for the same rate of flashing. He was comparing a star of one type with another star from the second family. Once this mistake was corrected, it was found that the Andromeda galaxy was nearly 2.5 million light years away. As we have seen, this makes it the most distant thing visible with the naked eye.

Hubble's discovery that the Milky Way with its billions of stars was just a small part of the universe would have been a remarkable achievement on its own – but he went on to discover something

The Andromeda galaxy seen from the Hubble telescope

even more amazing using a spectroscope. As we've seen, this instrument helps astronomers work out which elements are inside a star. In fact it's remarkable that it is easier to tell which elements make up a star than it is to discover how far away it is.

When you heat things up they start to glow. But rather than give off every single colour of the rainbow, different chemical elements produce light in specific bands of colour. The metal sodium, for example, has a distinctive yellow band, which is why sodium vapour street lamps give off that colour. But this process works in reverse as well. When white light passes through an object like a star that contains different chemical elements, those colours distinctive to

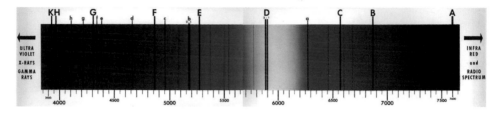

Dark lines in the spectrum of light from a star are signatures of the elements present in it

the elements are absorbed, leaving black gaps known as absorption lines in the spectrum. So if there's a black line in the yellow part of the light from a star corresponding to that key colour for sodium, you know that there is sodium present.

While Hubble was studying the colours coming from distant stars in the newly discovered galaxies, he noticed something strange. The colours seemed wrong. The lines were spaced out just as you might expect – but they appeared in the wrong part of the spectrum. They had shifted.

We've all come across the way a different sort of spectrum can shift. When you hear the siren on an ambulance or police car coming towards you, passing you and moving away, the sound distorts in a familiar fashion. It's called the Doppler Effect. The waves that make up the sound are squashed together as the vehicle approaches and stretched out as it moves away, which changes the pitch of the sound you hear.

We can't hear galaxies, but, if they are moving towards us or away from us, the colour of the light undergoes a Doppler shift. If you think of light as a wave, then the Doppler Effect makes the frequency of that light change – and we see the frequency of a light wave as its colour. If you prefer (as I do) to think of light being made up of particles called photons, then the photons coming from a galaxy heading towards us have extra energy, while those from

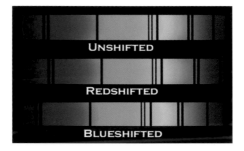

Movement of distant stars can be detected from the redshift or blueshift of the absorption lines

a galaxy heading away have less energy. We see the energy of photons as their colour.

Either way, when a galaxy moves towards us its light becomes more blue – it undergoes a blue shift. If a galaxy is moving away, then its light shifts towards the red and there's a red shift. Hubble discovered that almost every galaxy he observed was red shifted. Apart from a few local exceptions, like the Andromeda galaxy, which has a blue shift, all the galaxies are heading away from us. Even more extraordinarily, the further away a galaxy is, the faster it is moving.

What we are seeing here is two distinct effects. For (relatively) close structures like the Andromeda galaxy, there is going to be a sizeable gravitational attraction between it and the Milky Way. These galaxies are on a collision course. (There's nothing to worry about – we've about four billion years before they meet, by which time our Sun will probably have swallowed the Earth anyway.) But most of the galaxies were demonstrating something as awesome as it was unexpected. They are all red shifting, moving away from us. The universe as a whole is expanding.

It might seem strange that we appear to be at the centre of this expansion – because apart from a few anomalies like Andromeda, every galaxy is shooting away from ours in all directions. But what is really happening is rather different. It's not so much that the galaxies themselves are moving, but that the space in which they sit is expanding. This is difficult to get your head around, because it is happening in three dimensions at once. It's easier to visualize the process in two.

Think of a balloon (you can try this at home). The surface of the balloon is two dimensional – it has no depth. Blow the balloon up a little bit and draw dots all over it. Then blow the balloon up some more. Each dot will stay at the same point on the balloon. They don't move across the surface of the balloon because they are fixed in place. But because the balloon *itself* gets bigger, the dots move away from each other. Pick any dot and all the other dots will move away from that point. Similarly, pick any galaxy in the expanding

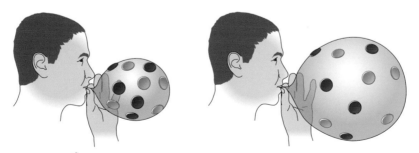

The expanding universe is like a balloon – pick any galaxy and the rest are moving away from it

universe and all the other galaxies are moving away from it. Despite appearances, the Milky Way doesn't occupy a special position.

There is a remarkable consequence of this expansion, which you can see if you imagine making a video of the universe (or our balloon) expanding and then run it backwards. If we think of the balloon again, as we run the movie in reverse, the balloon gets smaller and smaller. If it was a very special balloon that could get as small as you like, rather than going all limp and floppy, then at a particular point in time it would vanish away to a single point. Now imagine doing the same thing with the whole universe. If you track it backwards, about 13.7 billion years in the past, the universe disappears to a point. This is what is sometimes called the Big Bang.

You can even see where the Big Bang happened. Point at the space about 10 centimetres in front of your nose. This is where the Big Bang happened. This works wherever you are in the world. In fact, it works wherever you are in the universe. Because it is space itself that is expanding (or that shrinks as we look back in time), you can say that any point in the universe is where the Big Bang happened, because if you go back far enough, every point was in the same place. Every point in the universe, including the one 10 centimetres in front of your nose, emerged from the single point that was the Big Bang.

Once we have set an age for the universe, we can set a limit on how far we can see. If the universe formed 13.7 billion years ago, then the longest that light can have been travelling towards us is 13.7 billion years. It can hardly have set off before the universe began. You might think this means that the visible universe is 27.4 billion light years across (because light has been coming to us for 13.7 billion years in each direction), but that doesn't allow for the expansion we know has been taking place in the universe.

If light has been travelling for 13.7 billion years before it reaches us, that light set off 13.7 billion years ago. But while it has been on its way, the universe has expanded (a lot), so much in fact that most distant light shows us objects that are around 40 billion light years away. From this, we can say with confidence that the universe is at least 80 billion light years across. It may be bigger – it may even be infinite – but this is the limit of what we can see.

It is quite awe inspiring just how far our picture of the universe has changed since the original idea that the solar system and a surrounding sphere of stars was everything. We originally thought there was nothing else, just the Sun, a few planets and a hundred or so stars. With better equipment we believed that the universe was the Milky Way, which turns out to have around 300 billion stars

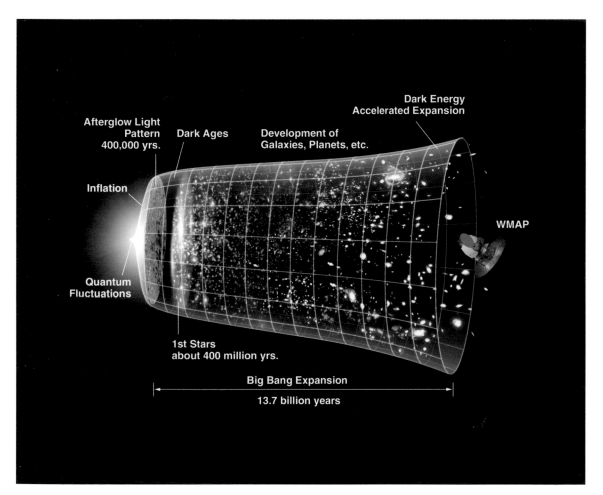

The universe from Big Bang to the present

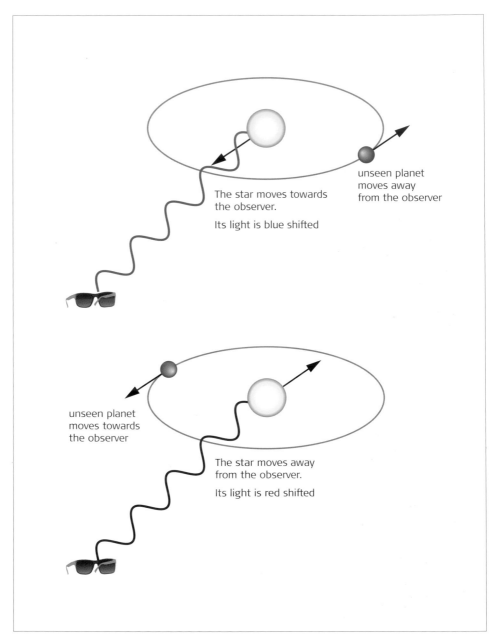

The star moves towards the observer.

Its light is blue shifted

unseen planet moves away from the observer

unseen planet moves towards the observer

The star moves away from the observer.

Its light is red shifted

Planetary motion causes a wobble in a star which can be detected as a blue or red shift in it. And now we know that there are other galaxies so vast that some have as many as 100 trillion stars in them. Not just a few galaxies, either. There could be as many as 150 billion galaxies in the part of the universe we can see.

Many of those stars are very different from our own. But if we take into account those that are relatively similar, there could be as many as 50 billion planets in the Milky Way alone. Until recently it has been impossible to tell if another star has planets. They are too far away to see in the reflected light of their star. But the motion of a planet around a star gives it a distinctive wobble.

By measuring the way that stars shimmy we can deduce what is in orbit around them. Over five hundred planets have been discovered outside the solar system so far. It's easier to find the big ones like Jupiter, which are unlikely to be inhabitable, because they have the biggest effect, but we have also discovered some smaller planets that could support life. The universe is, without doubt, a remarkable place.

chapter 6

A bang
or a whimper?

Once Hubble discovered that the universe was expanding, there was a shocking implication. Assuming this expansion had not suddenly started without reason, it seemed possible to run a mental movie of that unfolding universe backwards and discover the beginning of everything. This is the picture of the origin of the universe that we call the Big Bang, starting with the entire universe in a single point.

Although there are other ideas on how the universe began (we will meet them in Chapter 10), the Big Bang remains our best and most widely supported theory. So it seems right to take a step back from star gazing and travel through time to follow the lifecycle of the universe. As we do so, you should bear in mind the proviso that while the Big Bang theory fits all current observations, it had to be

An artist's impression of the distinctive jets of a quasar

modified to get that fit, and it is only one of several theories. For convenience I will describe the Big Bang as if it is fact, but (as is usual in science) it should be treated as the best available theory, not ultimate truth.

The Big Bang begins around 13.7 billion years ago with the entire universe in the form of an infinitely small point called a singularity. A lot of people wonder 'What came before the Big Bang?' and in the basic Big Bang model the answer is very simple. Nothing came before the Big Bang. Not only was there nothing before, there *was* no before. Because our understanding of the Big Bang fits within a model of the universe that is based on Einstein's theory of general relativity.

General relativity was one of the masterpieces of twentieth-century science. It describes how gravity acts. Back in the seventeenth century, Isaac Newton had described the basic law of gravitation, but he had not attempted to explain how gravity manages to pull things about from a distance. In the Latin original of his masterpiece *Principia*, he wrote 'Hypothesis non fingo' – I frame no hypothesis.

General relativity does much more than account for gravity. It is a description of how space and time behave under the influence of mass – what stuff does to them. In this, it doesn't treat space and time as separate entities. Instead they are considered part of a single system called spacetime, where time is another, rather special, dimension. And in the Big Bang model, spacetime begins with the Big Bang. The beginning of spacetime is the beginning of everything, including time itself. There can't be a 'before the Big Bang' because there was no time.

Einstein's general relativity explains gravity as a warp in space and time

In this picture, science has no mechanism for explaining why the Big Bang took place. It just happened. You can say 'God did it', or 'It was spontaneous', or 'We don't understand why it happened' – the outcome is the same. When we look at other possibilities for the way the universe began in Chapter 10 we will see versions of the Big Bang that did have a cause, but in the standard Big

The Big Bang is thought is thought to have begun from a point singularity

Bang model it is impossible to identify a reason, the universe just came into being as an infinitesimal speck.

In a tiny amount of time this ridiculously small entity began to grow. The very beginning, when the energy involved is effectively infinite, is outside scientific theory, but once that expansion began, we have something we can apply science to. It's still a very small something, much smaller than an atom. For many people this seems puzzling. How can it be that everything we now see in the universe – every star and planet, every bit of matter in every one of billions of galaxies – was compressed down to such a tiny size?

Admittedly, atoms are mostly empty space. By far the biggest component of your body is nothing. If you magnified an atom (any atom) until you could see the nucleus – the central part of the atom which has most of its mass – that nucleus would be about the size of a fly in an atom that was, relatively speaking, the size of a cathedral. Apart from a few electrons zipping around the outside of the atom in a cloud of probability, the rest of it is emptiness. But even if there were some way in those first beginnings of the universe to do away with the empty space we have a problem.

It has been estimated that if you took the entire human race and somehow removed all the space from their atoms (super villains should take note – this isn't physically possible) they would be compressed down to about the size of a sugar cube. But we are talking about fitting what will become the entire universe – all those stars and galaxies – into a space much smaller than an atom. Taken at face value, this just doesn't seem right.

However, there are two important factors that make that cosmic seed possible. One is that the early universe would not have had any matter in it whatsoever. Einstein's other great work, the special theory of relativity, included the best known equation of all time: $E=mc^2$. This says that energy (E) can be converted into mass (m) and mass into energy. There's an awful lot of energy in matter. That 'c^2' bit is the square of the speed of light, which is a very big number. But energy and matter are interchangeable. And energy doesn't take up space the way matter does.

The other contributory factor to squeezing the entire universe into a point is that, bizarrely, the universe really doesn't have to have started with much in it to end up the way it is. This is because gravity can be considered as a kind of negative energy. Once you combine the mass of everything in the universe with the gravitational pull that all matter provides, they pretty well cancel each other out. There didn't have to be a huge amount of content when the universe was first formed.

The first significant event happened pretty early on. Around 10^{-36} seconds after the beginning, the universe is thought to have gone

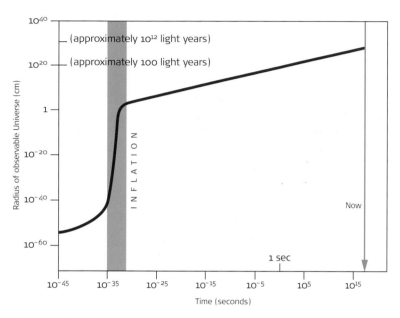

The period of inflation produced a vast expansion in the size of the universe

through a brief but essential phase called inflation. This is expansion, just as we currently see the universe to be expanding, but at a much faster – a phenomenally fast – rate. If you aren't familiar with scientific notation, the start time of inflation at 10^{-36} seconds may not be particularly meaningful. The minus sign before the 36 means 1 divided by 10^{36}. And 10^{36} is 1 with 36 zeroes after it. So this inflation took place when the universe was just 1/1,000,000,000,000,000,000,000,000,000,000,000,000 of a second old.

Inflation wasn't a long process, but the growth was incredible. By the time the period of inflation finished, the universe was still only around 10^{-32} seconds old. This epic expansion took place in a ludicrously small fraction of a second. Yet in that time, the universe grew by at least a factor of 10^{30} and possibly as much as 10^{70}. Take those figures in slowly. The universe was at least 1 with 30 zeros after it bigger than when it first started. We're still not talking about a huge universe – it was smaller than a present day galaxy and quite possibly only the size of a grapefruit – but it had been vastly smaller before. To expand so quickly, the universe would have to have stretched and enlarged far faster than the speed of light.

There seems to be a problem here, because Einstein's special relativity also tells us it's impossible for anything to travel through space faster than the speed of light in a vacuum (around 300,000

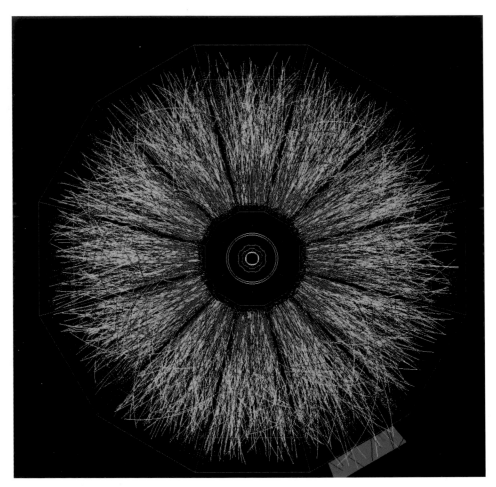

Visualization of a quark gluon plasma created in a collider

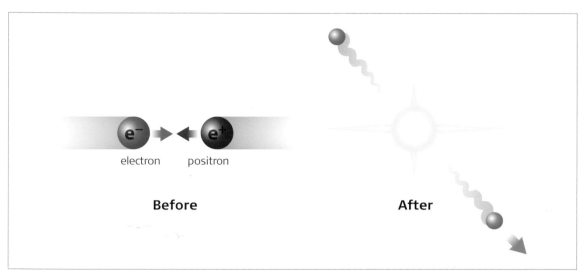

Matter and antimatter anihilate to produce energy in the form of light

kilometres per second) – but we have to remember that when the universe is expanding, whether the 'normal' expansion that gives galaxies their red shift or during inflation, things aren't moving in space. It is space itself that is expanding, and there is no limit to the speed at which that can happen.

When the universe was just a fraction of a second old, it was so hot that the particles of matter that began to form from the pure energy were not atoms, or even the familiar subatomic particles like protons, neutrons and electrons, but instead were quarks. These are smaller particles that make up protons and neutrons – each of which is composed of three quarks. We don't see quarks now because it requires a vast quantity of energy to break a proton or neutron apart, but in the sub-second universe these mysterious particles were the natural form for matter, forming what is known as a quark gluon plasma.

By the time the universe was about a second old it had cooled enough (which basically means the individual particles had lost energy as it expanded) for protons and neutrons to form. But these were not just familiar matter – there was also antimatter. When energy is converted into matter, it usually forms pairs of particles, matter and antimatter. Although antimatter sounds like something out of science fiction, it is real enough. An antimatter particle is just one where some key properties, notably the particle's charge, are reversed.

Probably the best known antimatter particle is the positron. This is an antielectron and is identical to an electron in pretty well every measure, like mass and the size of its charge, but the positron

Electron and positron come together to annihilate

has a positive charge where the electron is negative. If matter and its equivalent antimatter particle collide, they annihilate each other, producing a burst of energy and potentially other types of particle.

Very soon most of the protons and neutrons combined with the antimatter equivalent, reverting to energy but also producing electrons and positrons (and other related particles). Within a few seconds these too had wiped each other out. Yet there seems an oddity in the symmetry between matter and antimatter that means,

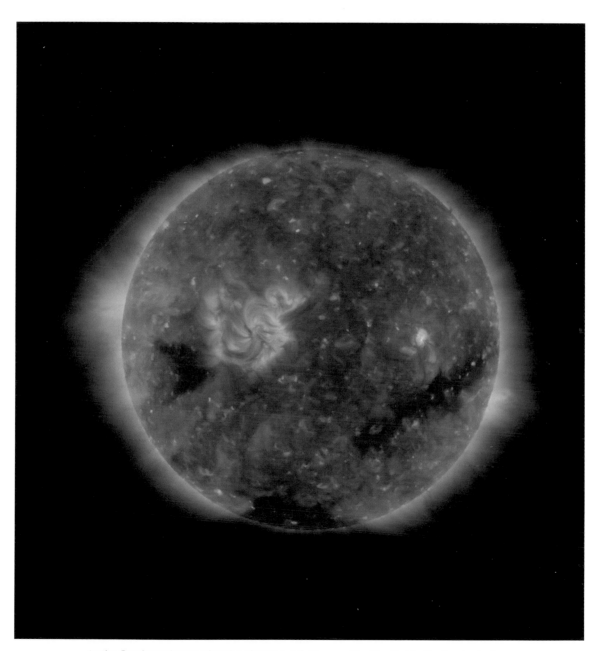

In the first few minutes after the Big Bang, a fusion reaction like that in the Sun took place

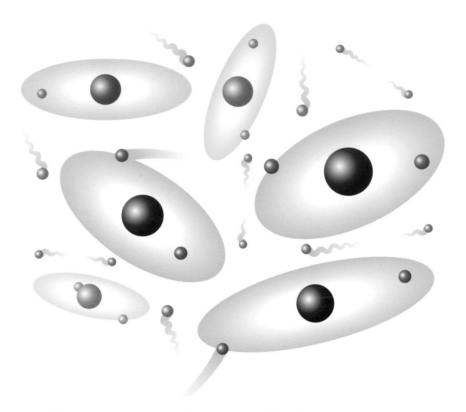

After around 370,000 years atoms formed and the universe became transparent

in each case, some of the matter particles would be left over. It's not clear if this effect is large enough to explain the amount of matter we see – this is one of the weaker aspects of the Big Bang story – but we know that don't now see lots of antimatter in the universe, so something must have happened to change the balance.

At this stage we're about three minutes into the life of the universe, and it is mostly made up of energy in the form of photons of light – but there is also matter, and for the next few minutes the universe has enough temperature and pressure to act like a massive star, converting hydrogen ions (a hydrogen ion is a hydrogen atom without an electron, which is just a single proton) into helium ions, just as the Sun does today. This fusion process only lasted around a handful of minutes. By then the temperature had dropped from around one billion degrees Celsius to a mere 10 million degrees. The expansion of the universe had taken it beyond the limits required for the process that powers stars to continue.

By now the universe contained hydrogen, helium and a small amount of the next element up in weight, lithium. For the moment, all this matter was in the form of ions – positively charged particles, with electrons flashing around separately. This is what's called plasma. The charge in the plasma makes it difficult for photons of light to travel far, because they interact easily with charged particles, so the universe was opaque.

Following the end of the star-like period, a fair amount of time elapsed as the universe expanded and cooled. After around 370,000 years, those hydrogen, helium and lithium ions had lost enough energy to make it relatively easy for electrons to hook up with them, turning them from ions into the atoms of the elements themselves. Now that proper atoms had formed, the universe became transparent. No longer were photons of light unable to travel more than a tiny distance before being captured. They began to flow freely, making this a significant point in time, which we will come back to in the next chapter.

At this stage, thanks to that period of incredibly quick inflation, the universe was evenly spread with matter, with just tiny variations in density caused by the fluctuations that naturally occur at the level of individual particles, which obey quantum theory, predicting a randomness in behaviour. But, over time, those tiny variations in the density of the universe began to grow. If all the matter in the universe had been spread out perfectly evenly then the universe would remain stable, but where there were little clusters of matter, other particles were attracted by the slightly larger gravitational pull.

More time passed. Between a hundred million and a billion years into the life of the universe, these clumps of matter had got so big that they formed stars and the early beginnings of galaxies, structures that we now call quasars (more on these in Chapter 9).

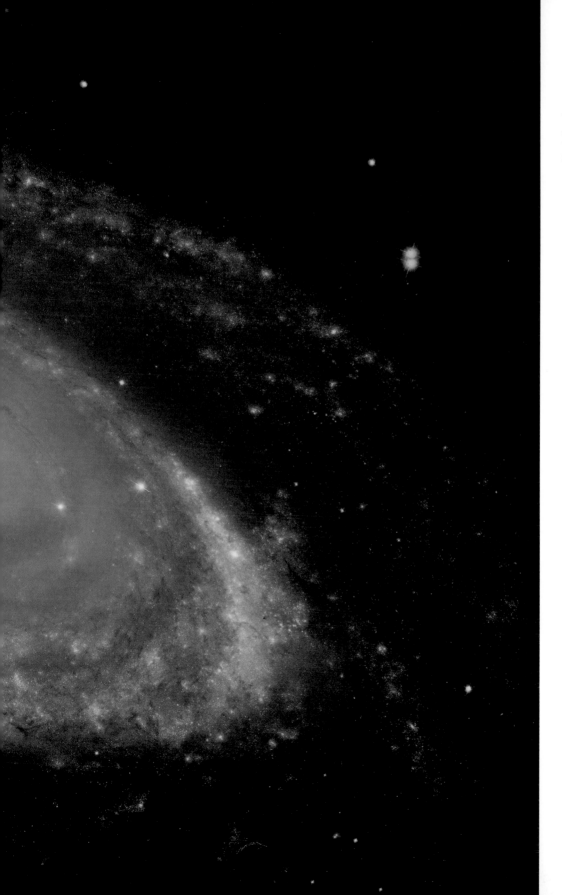

The spiral galaxy M51, similar to the Milky Way in structure

In a vast, evolutionary timescale the forms of galaxies and the stars we now know began to coalesce over billions of years.

In the earliest stars, as hydrogen fused to form helium and helium fused into heavier ions, the whole range of elements up to iron were produced. This is as far as a conventional star can go because iron is too stable to fuse to form anything else. But some stars, at the end of their lives, exploded in vast supernovae, producing so much energy that they were able to turn out a much wider range of elements, all the way up to the biggest natural atom, uranium. In the stars and supernovae, every bit of matter other than the hydrogen, helium and lithium from the Big Bang was formed. This means that the stuff that makes up you (and your car and the Earth and everything around us) came either from an early star or from the ultimate beginning of the universe. Every bit of us is billions of years old, and much of it is stardust.

Not all galaxies and stars formed at the same time. The Milky Way was not one of the earliest galaxies – it seems to have begun forming around 8.3 billion years ago when the universe was already over five billion years old. But the process of formation is a continuous one, driven by the insistent force of gravity. Our solar system, for example, only began to come together around five billion years ago, forming from material from those earlier generations of stars as well as the remnants of the Big Bang that are spread throughout space. By 4.5 billion years ago the basic structure of the solar system we know today was in place.

Expansion continues at an increasing rate. The universe continues to cool as it has for the past 13.7 billion years. Stars still form and stars still explode. The evolution of the universe is a process that is still underway as we watch the skies around us.

Another spiral galaxy, M81, sometimes known as Bode's galaxy

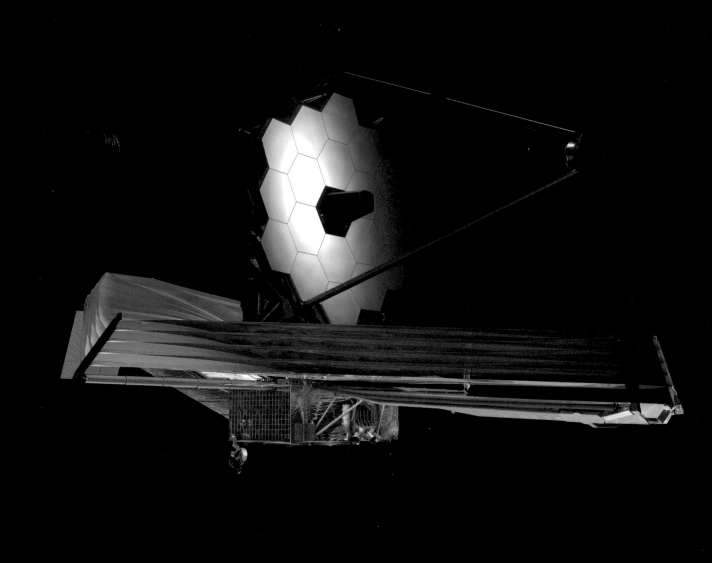

chapter 7

How do we know?

In the previous chapter we made bold claims about the history of the universe, but how do cosmologists know what happened billions of years ago? For that matter, how is it possible to work out what's going on in a star many light years away? Astronomy and, particularly, cosmology, the science of the universe as a whole, is highly restricted by practical limitations. We can't do experiments. We can't go out and touch. But we can make some remarkable deductions from the evidence that reaches us across space and timeWe've already seen that light provides an essential connection to the rest of the universe. Travelling at 300,000 kilometres per second, photons of light can cover vast distances. Yet because of the time that light takes to travel, we see different parts of universe

A mock up of the James Webb space telescope

The twin Keck telescopes with vast 10 metre mirrors

at different stages of their development. When we observe the Sun, things are fairly immediate. We see our neighbourhood star as it was eight minutes ago. If there was some way to make the Sun disappear instantly, we wouldn't know for eight minutes and our lives would go on unchanged for that brief time. Gravity also travels at the speed of light, so we would continue to see the Sun and be kept stable by its gravitational pull for eight minutes after it disappeared.

As we've also seen, our next nearest neighbour, the star Proxima Centauri, appears in the sky as it was a little over four years ago, and the Andromeda Galaxy that we can just about see with the naked eye is the way it looked around 2.5 million years ago. But we can see much further than that. Whether the target is distant galaxies or strange phenomena like quasars (see Chapter 10), modern telescopes enable us to peer back into the ancient history of the universe and its evolution.

In principle, with a powerful enough telescope, the furthest we can look back is around 13.4 billion years to the point where the universe became transparent. The light from before then is lost, but the light that was zipping around the universe when it became transparent is still out there. At the time it would have consisted of extremely high energy photons of light – gamma rays. But something has happened over the billions of years since that light was emitted: The universe has expanded.

Remember that it is space itself that is growing bigger. As it does it will have a similar effect on the light that is passing through it to the way that light from a distant galaxy is affected. It will be red shifted. And the longer that the light has been in motion, the more the universe will have expanded since it set off, resulting in more pronounced red shifts.

What was very high energy light has gradually moved through X-rays, ultra-violet, visible light and infra-red all the way down to microwaves. These are a kind of light where photons have much lower energy than the visible – microwaves have more in common with radio than visible light – but we are familiar with microwaves because they have just the right energy to get water molecules

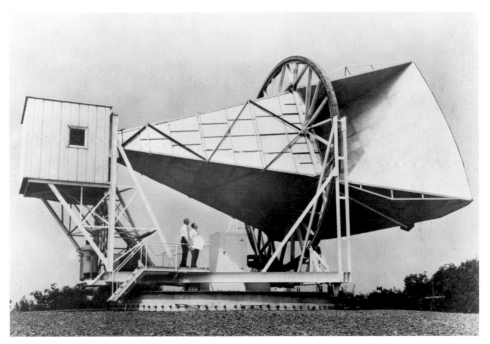

The Bell Labs Holmdel site that first identified the cosmic microwave background radiation

excited. This makes them great for heating up an item with water in it, like a piece of food, hence the use of microwaves in an oven.

Back in 1965, two researchers at the Bell Labs facility in Holmdel, New Jersey, were trying to use an aerial to pick up signals from the early Telstar communication satellite to continue their astronomy. It had been discovered by then that stars didn't just give off visible light, but also a wide range of the electromagnetic spectrum, including radio. The researchers, Wilson and Penzias, were looking for radio output from the edges of the Milky Way, but instead they found a strangely uniform background hiss that came from every direction.

It was similar to the static that was picked up on old analogue TVs if they were tuned between stations – in fact part of that visual static and hiss was exactly the same signal that Wilson and Penzias received. For a while, the radio astronomers thought they were receiving earthbound interference. This can be a common problem with radio astronomy – a faulty motor in a vacuum cleaner a few miles away can produce a false signal. But with careful analysis they showed that it wasn't coming from any of their equipment, and it was just as strong whichever direction they pointed their antenna.

Another suspect for the cause of this mysterious signal was the droppings that were building up because a family of pigeons was perching on the wide horn of the telescope. But even when they got rid of the pigeons and cleaned up the metal surfaces the hiss remained. It was only when they talked to some other scientists who were looking for such a signal that they realized what it was they were picking up. The microwaves they were the receiving were the remnants of the light that had been set free when the universe became transparent around 370,000 years after its birth.

This 'cosmic microwave background radiation' has been called the echo of the Big Bang. This is rather flowery and inaccurate language, considering it is the remnant of something that took place a third of a million years after the Big Bang, but it gives us an insight into the earliest view of the universe we can have. This radiation takes us back as far as light lets us see. In principle we could go one step further, though. At around one second after the

A neutrino image of the Sun

Big Bang the universe became transparent to a different type of particle, the neutrino.

Neutrinos are produced in vast quantities by nuclear reactions. The Sun, for instance, pours out huge numbers of neutrinos. Around 50 trillion of them pass through your body every second – but they are so bad at interacting with other particles that we don't notice them, nor can we directly detect them. Neutrinos were dreamed up to explain some missing energy in nuclear reactions back in 1930, but it wasn't until 1956 that one was detected by spotting the outcome of a very rare collision between a neutrino and another particle.

Neutrino detector
at the Japanese
Kamioka
Observatory

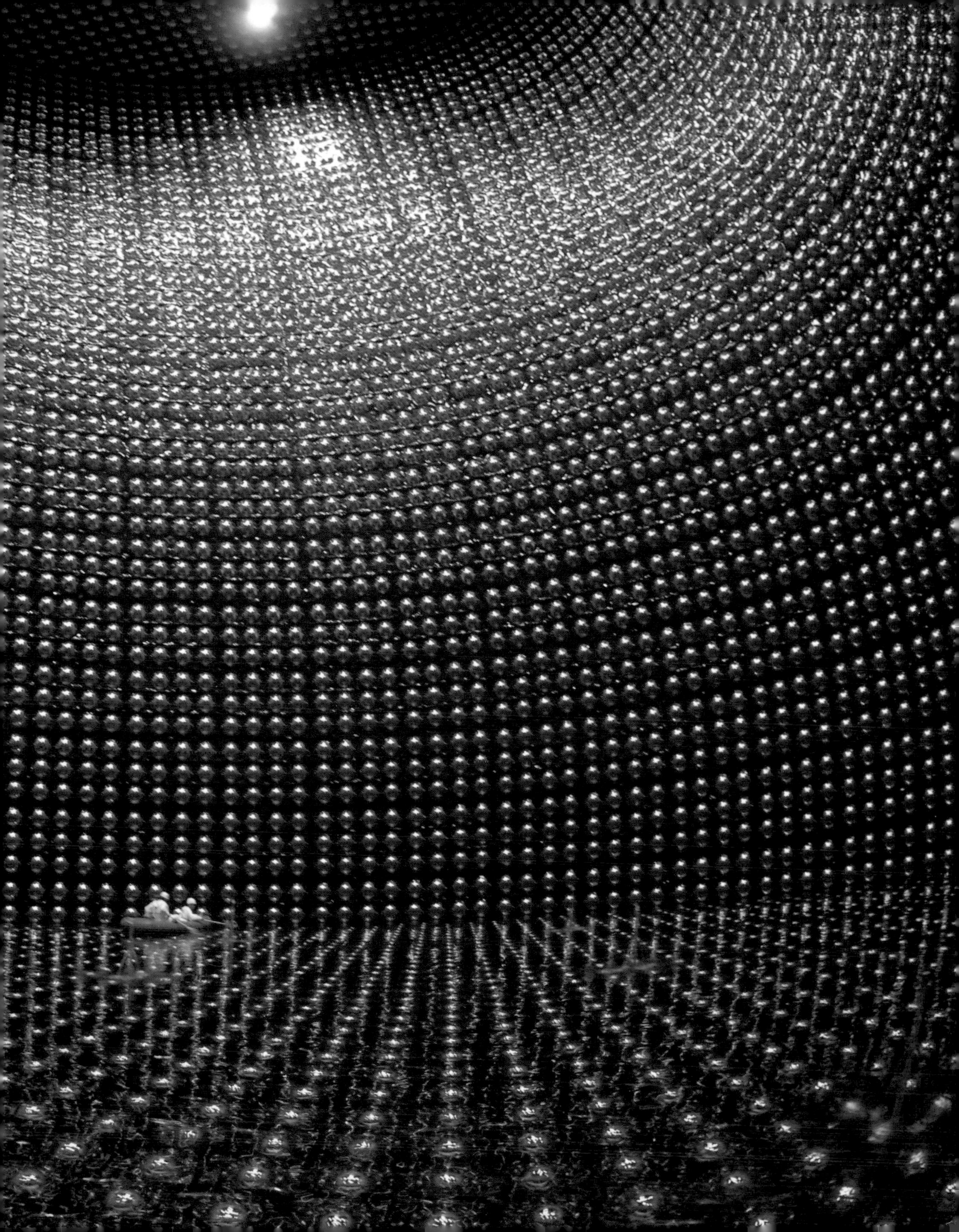

Neutrino detectors are usually buried far underground in mines. They contain large quantities of material, often as mundane as cleaning fluid, which is then monitored by an array of detectors for new particles. As practically all the other particles and radiation from the outside will be stopped by miles of rock, the best guess is that when a particle breaks up to give off energy and other particles, then the cause of those reactions must be the neutrinos that are constantly sweeping through the Earth.

A neutrino detector has even been used to make a crude image of the Sun – the first neutrino telescope. As if to demonstrate just how little neutrinos care for ordinary matter, the picture was taken through the Earth, with the Sun on the far side from the detector. If we can improve our neutrino detectors, we might be able to detect this cosmic neutrino background radiation from the first second of the universe's existence. For the moment, though, we are limited to the view that microwaves give us from 370,000 years after the Big Bang.

When detectors on Earth are brought into play, like the crude radio telescope used by Wilson and Penzias, the cosmic microwave background is extremely smooth, producing identical levels from every direction. It is one of the reasons why it was identified as the remnant of the Big Bang because that was expected to be the same from every direction. But in more recent times we have been able to study the cosmic background radiation in a lot more detail and have uncovered minute variations.

This new view of the radiation pattern is down to two satellites, COBE and WMAP. COBE dates back to 1989, while WMAP was launched in 2001. What these satellites were able to do with increasing accuracy was to plot tiny variations in the cosmic microwave background. The result looks dramatic in their images but the contrast is hugely amplified. The actual variation is around 1 in 100,000, which represents only tiny changes from the constant background level.

When you look at the stretched-egg-shaped images from WMAP it is hard to make out just what you are seeing. It is thought that the pattern in the radiation is the result of the tiny variations in the

Herschel's largest telescope at Slough, near Windsor

The cosmic microwave background view of the universe from the WMAP satellite

LEFT *The auxiliary telescopes at Cerro Paranal in Chile*

BELOW LEFT *Mount Palomar's Hale telescope, for decades the largest in the world*

BELOW *The giant Arecibo radio telescope in Puerto Rico*

makeup of the universe that would soon result in galaxies forming. If this is correct, what we see is the equivalent of an ultrasound scan of the embryonic universe – surely a remarkable picture.

The cosmic microwave background radiation takes us back to the earliest visible moments of the universe, but since the Andromeda galaxy was first spotted we have been looking further and further into space by more direct means. The first telescopes used lenses or mirrors to focus light from distant objects in the sky, making them visible for the first time. Although we still use these optical telescopes, they are now a part of a huge battery of telescopic devices, taking in the whole span of the electromagnetic spectrum from radio waves to gamma rays.

The point of using these different frequencies or photon energies is that each body in the universe radiates light in its own particular pattern of energy. Some are clearer in visible light. Others are practically invisible optically, but shine in radio or X-rays. There are also materials in the universe like dust that absorb certain energies of light better than others. By observing using the widest range of light we get the best images.

Some of the great telescopes are, as they always have been, based on Earth. Originally such telescopes were used in the cities where the early astronomers lived. Galileo, for instance, first demonstrated his telescope in Venice, while Herschel discovered Uranus from his back garden in the British city of Bath, and later built much larger telescopes in Slough (now an industrial town, but then a village on the outskirts of Windsor, handy for Herschel's royal patron in his home at Windsor Castle). But there were increasing difficulties in getting a good image in such places.

With the industrial revolution, the levels of soot and other pollutants in the air increased. Coal fires threw out millions of tiny particles, while the heat from large buildings made the air ripple, distorting a telescope's view. At the same time, street lights and other artificial lights at night increasingly compromised the darkness that was essential for good observation. The new genera-tion of telescopes, including the one that Hubble would use to discover how far away the Andromeda galaxy was, were built on mountaintops in deserts where the air was thin, steady and unpolluted by a city's streetlights.

The thinness of the air on those mountaintops was important. Air itself is a constant irritation for astronomers. Of course they have to breathe like the rest of us, but air scatters light. This is obvious whenever we look at the blue sky. If the air didn't scatter the light we would see a black sky with a brilliant sun, but the molecules in the air send the light flying in all directions, carrying it across the sky to provide a uniform illumination. (Blue light is scattered more than red, hence the blue sky.)

The effect of this scattering is to make images fuzzier than they could be, and that assumes steady air. Add some heat and the air begins to shift and ripple, distorting viewing even further. When Hubble discovered the expansion of the universe, he was working on Mount Wilson near Pasadena in the USA. The biggest telescope there at the time was a world-beater. The mirror that was used to collect light was 2.5 metres (100 inches) across.

This record would be put in the shade in the 1940s by the mighty Hale telescope on Mount Palomar in San Diego County. At five metres (200 inches) it was the most powerful instrument in the world until 1976. The mirror is made from 65 tonnes of glass supported on a 500-tonne mount. Although it remains one of the world's most famous telescopes, it has now been bettered many times over.

At the time of writing, the world's biggest telescope is the Gran Telescopio on the Canary Islands, with a segmented mirror 10.4 metres (409 inches) across. (The mirror is divided into a series of hexagonal segments, as it would distort under its own weight if it were a single piece.) Close behind are the two Keck telescopes on Mauna Kea, Hawaii, at 10 metres (394 inches). The biggest single-mirror telescopes form the large binocular telescope on Mount Graham, Arizona, with two 8.4-metre (331 inch) mirrors in parallel, giving it the equivalent resolution of a 23-metre mirror. This is closely followed by the 8.3-metre (327 inch) Subaru telescope on Mauna Kea and the remarkable Very Large Telescope (VLT) at Cerro Paranal in Chile, which has four 8.2-metre (323 inch) mirrors.

These are extraordinary devices. The Keck telescopes, acting together, for example, could distinguish the headlights of a car 16,000 miles away. But there are limits to what any telescope based on Earth, however remarkable, can achieve. There are tricks that can be used to get round some of the distortions caused by the atmosphere. Adaptive optics, for example, monitor the image produced by the telescope and change the shape of the mirror to counteract distortions. But there will always be some scattering, some fuzziness. And even remote, dry locations sometimes suffer from bad weather. What's more, wherever an observatory is sited, it can only be used during the hours of darkness. But a new generation of telescopes has changed all this: telescopes mounted on satellites.

The name we all know is the Hubble Space Telescope (HST), which has produced so many wonderful photographs of everything from planets in the solar system to distant galaxies. This is the both the workhorse and the celebrity of space telescopes, although its career nearly ended before it had begun. The HST was launched in 1990 from the shuttle *Discovery*, but the images it produced were terrible. After frantic diagnostics, it was discovered that the mirror was the wrong shape.

The amount of distortion responsible for the fault demonstrates just how critical manufacturing is on this kind of scientific instrument. The total error on the Hubble mirror was around one fiftieth of the width of a human hair – but that was enough to make the images it produced all but useless.

In 1993, the shuttle *Endeavour* drew alongside the telescope for a repair mission. It was simply not practical to replace the mirror while the satellite was in space – it was too large a component. And there was no way the astronauts could reshape the mirror. It still has the fault to this day. Instead, over the two-week mission, the astronauts replaced the optics that handle the light from the mirror with systems that had been carefully tweaked to have exactly the same flaw in the opposite direction.

Repairing the Hubble telescope was the large-scale (and extremely expensive) equivalent of what happens every day in

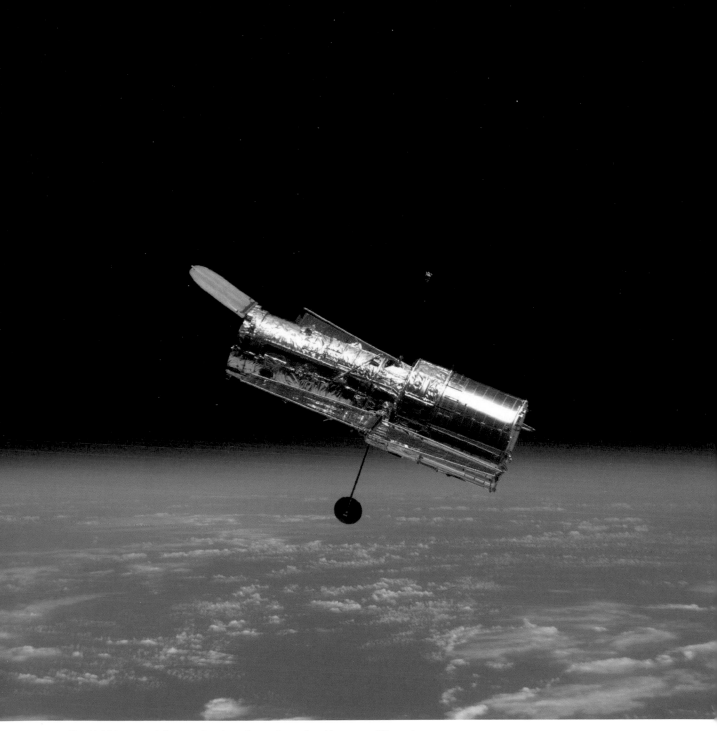

The Hubble space telescope has transformed our visual imagery of the universe

opticians the world over. They were fitting the telescope with their version of correcting spectacles. And the result was spectacular. Even though the HST's mirror is smaller than the one Edwin Hubble himself used on Mount Wilson, the satellite immediately began to produce images of unparalleled clarity and detail.

There are now plenty of telescopes in space dealing with different parts of the electromagnetic spectrum, like infra-red, ultra-violet and X-rays. But from 2013, if the budget survives cost cutting, the HST will be eclipsed in the visible light field by a new satellite, the James Webb Space Telescope, named after NASA's second administrator. The Webb will have a much larger mirror – 6.5 metres (256 inches) across – made from 18 separate segments coated in 24-carat gold.

The Very Large Array radio telescope in New Mexico

Meanwhile, back on Earth, we have also been looking beyond the visible. Since the 1960s and the re-use of the Telstar aerial as a telescope, radio astronomy has come on immensely and has proved a hugely effective tool in understanding what is happening in the universe in more detail. Radio photons are much lower energy than visible light (thought of as a wave, the radio waves have a much longer wavelength). This means that the images produced by radio telescopes are less distinct, less able show detail. But in exchange for loss of detail, radio can detect sources that are difficult to see or completely invisible in the optical range.

It is also much easier to build a massive radio dish than it is to build a big mirror. Radio telescopes can easily reach 75 metres (250 feet across), whilst the biggest at Aricebo in Puerto Rico is a fixed dish a thousand feet across. This is too big to be steered so it can only pick up signals from where the Earth happens to be pointing. With modern computing technology, these fixed behemoths have become increasingly redundant. More recent radio telescopes work instead by having large arrays of relatively small moveable dishes. Computing power enables the signal to be woven together to provide a single image equivalent to that of an impossibly large dish.

These different technologies have shown us a lot of what is out in the universe. Yet as astronomers have discovered more about galaxies and the other structures out in the vastness of space, they have reached a shocking conclusion. Most of the universe appears to be missing.

chapter 8

Most of our universe is missing

In the previous chapter we found out more about the part of universe we can see – but this is only a tiny fraction of reality. It's not surprising that we can't find everything that's out there. Only stars and related phenomena give off light. Anything cool, from a speck of dust to a whole planet, is dark and can't be picked up by telescopes, with exception of objects in our own solar system. We can only see these because the Sun's bright light picks them out. But even taking into account the limitation in our view of distant objects, astronomers in the mid-twentieth century began to realize that something was missing. Galaxies seemed to have more mass than could possibly be provided by stars and planets.

The Tadpole galaxy, around 420 million light years away, distorted by a neighbouring galaxy

The colliding spiral galaxies in Virgo

There seems to be a piece missing from the jigsaw. How could we know that something we can't detect is missing? The clue comes from the way things spin around. Just about everything in the universe seems to spin. Planets like the Earth spin as they rotate in their orbit round the Sun. The Sun spins too. Our whole galaxy, the Milky Way, rotates in a stately revolution. Spinning is part of the nature of the universe. But between the 1930s and 1960s observations were made that suggested some spins were wrong. Galaxies appeared to be spinning too fast for their size. When you get clusters of galaxies rotating around each other, they are also spinning too quickly.

Think of a lump of clay on a potter's wheel. Turn it at normal speeds and the clay will stay in place, ready to be worked. But spin the wheel fast enough and bits of clay will fly off, where the force applied to them by their movement is bigger than the force holding them to the main body of the clay. Spin the wheel faster still and the whole lump of clay will be flung off the wheel. The impact of the spin, known as angular momentum, is just too great for the stickiness that holds the clay in place.

When it comes to galaxies, the thing that's holding them together is not the stickiness of clay (which is ultimately down to the electromagnetic forces between the atoms in the clay) but gravity – and that too has its limits. If you imagine taking a galaxy and somehow cranking up the speed, turning it faster and faster, stars should start spraying out into the void like sparks from a pinwheel firework. The galaxies appear to be spinning so quickly that there shouldn't be enough matter for gravity to keep them together. They ought to be breaking up. But they aren't.

There are two possible answers to this conundrum. Either there is more mass than we thought in these galaxies, or the laws of physics that apply to everyday objects don't work in the same way with something the size of a galaxy. This second possibility isn't as unlikely as it sounds. We know that very small things like atoms behave totally differently from the objects we are used to seeing and handling. Galaxies are as different in scale to ordinary everyday objects as atoms. It is entirely possible that Newton's laws of motion and gravity (modified by Einstein's relativity) act differently on bodies the size of a galaxy.

This idea, of changing the rules for the way bodies on the scale of galaxies move, called Modified Newtonian Dynamics or MOND for short, is one explanation for the way that vast bodies rotate so quickly. But it's not the most popular suggestion. Instead, the rotation is usually blamed on dark matter.

Dark matter is all the stuff that we can't see in the universe, anything that has mass but doesn't give off light in any part of the electromagnetic spectrum. When we come up with a figure for the mass of a whole galaxy there is a lot of estimation going on. We can't check the weight of each and every one of billions of stars. But a reasonable approximation can still be made. Add in the assumed dust and planetary bodies and you are still way too short. Even assuming that there is the expected quota of black holes out there (more on these in the next chapter) the mass falls far short of expected values.

In fact it's estimated that if the odd behaviour of galaxies and galactic clusters is down to extra mass rather than MOND, we are looking for six times as much mass as there is in every star and planet, plus every other known dark object. This missing mass, the dark matter, accounts for around 25 percent of the entire universe, where just five percent is the matter and energy we can see and understand.

If dark matter exists, no one is certain what it's made of. There's a suggestion that it is made of 'weakly interacting massive particles' or WIMPs. This kind of dark matter would be a bit like sumo wrestler versions of neutrinos (see page 129). Like neutrinos, the WIMPs would have very little interaction with ordinary particles, but unlike neutrinos which have hardly any mass, and possibly no mass at all, these would be heavy particles, contributing a great deal to the overall mass of a galaxy.

The main alternatives to WIMPs have a related nickname – MACHOs. This stands for massive compact halo objects. These are collections of heavy, dark objects around the outside of a galaxy (in the 'galactic halo'). MACHOs would be made of ordinary matter

– dust, rocks, or even black holes – clustered around the outside of a galaxy which should give it that extra ability to spin. However, WIMPs tend to be favoured of the two, as there should be more evidence for the existence of MACHOs than is picked up.

The need for extra dark matter or MOND is emphasized by the results from the COBE and WMAP satellites. As we've seen, the faint pattern present in the cosmic microwave background seems to reflect where matter should eventually accumulate to form galaxies. But the amount of matter that the pattern suggests should be there simply isn't enough to form the universe as we see it now – if the analysis is correct, we need a large quantity of dark matter, or the influence of modified dynamics, to enable the galaxies to pull themselves together.

If dark matter is involved, WIMPs are favoured in this scenario.

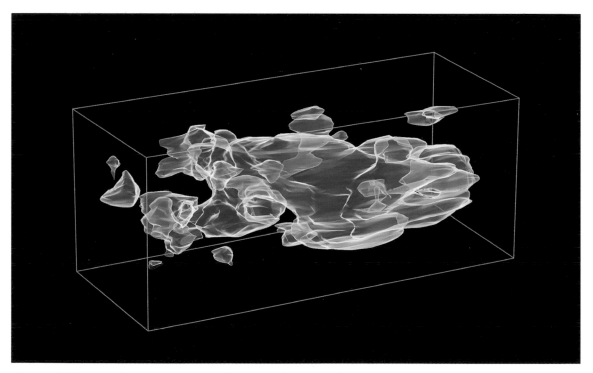

3D map of dark matter in a segment of sky: left is closest (present day), right furthest (early universe)
FOLLOWING PAGES *Artist's impression of dark matter objects*

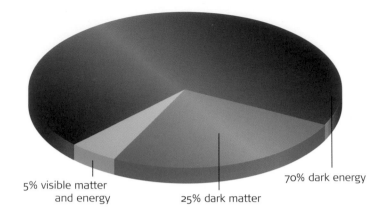

Only 4 to 5 percent of the universe is ordinary matter and energy

5% visible matter and energy

25% dark matter

70% dark energy

This is because of the low level of interaction WIMPs would have with other particles, other than gravitational attraction. It is thought that ordinary matter (dark or otherwise) would have been battered around too much by the powerful radiation that flooded the early universe. WIMP dark matter would be unaffected by this radiation, so would start condensing under the pull of gravity earlier, forming an invisible nucleus around which an ordinary matter galaxy could form as radiation energy levels reduced.

The existence of dark matter is an impressive piece of speculation, but you might wonder if cosmologists are any good at basic arithmetic. I mentioned that around 25 percent of the universe's content is attributed to dark matter, with another five percent accounting for all the visible matter and energy. That only adds up to 30 percent, so where is the remaining 70 percent? Remarkably, the rest of the universe is in a form for which we have no real explanation. We only have a name for it: Dark energy.

Bear in mind that mass and energy are interchangeable. Mass can be converted into energy, energy into mass. So when considering the total content of the universe we have to consider both. Dark energy is a wild energy assumed to be running throughout the universe that is responsible for a strange feature of the whole of space that we have already met.

We know from the observations of red shift in distant galaxies (work begun by Hubble) that our universe is expanding. We would expect one of three possible future outcomes. It's possible that things could continue the same way forever with a steady rate of expansion. It's possible that the expansion could gradually slow down, never actually reaching a stop, but getting slower and slower. Or the expansion could slow down, halt and reverse, so that the universe then began to shrink back to the opposite of a Big Bang, sometimes called a Big Crunch.

Some hoped for the third option, giving the possibility of an oscillating universe that first expanded and then contracted in an eternal cycle. The favourite was that expansion would gradually slow, but never reach a stop. However, it now appears that *none* of these scenarios applies. The expansion of the universe is actually accelerating.

Our universe is not only getting bigger minute by minute, but the speed at which it grows is increasing. And one thing we know about acceleration is that it takes force to make it happen. We need to apply energy to drive this acceleration. And the mysterious force that appears to be driving the expansion of the universe is called dark energy. Given the size of space and the rate of expansion it is experiencing, there has to be so much dark energy that it accounts for 70 percent of the entire universe.

Some have suggested that dark energy is simply a reflection of the natural energy of empty space. We tend to think that a total vacuum – no matter, no photons – would also have no energy, but quantum physics doesn't allow this. It tells us that there is a minimum possible energy the empty vacuum can have, which is purely dependent on the amount of space available.

This energy inherent to empty space will spend its time flipping between pure energy and virtual particles. These are pairs of particles, one matter and one antimatter, that aren't normally seen because as soon as they are created and before they can interact with anything, they annihilate each other and revert to energy. Although we don't usually see these particles, we know they exist as they can generate effects on nearby matter, or even be forced into real existence if one of the matter/antimatter pairs reacts with something else before they can recombine.

As the universe expands there is more space available (remember, it's space itself that is expanding). As this vacuum energy level is directly connected to the amount of space, there will be more vacuum energy, which it is thought may be responsible for the acceleration of the universe's expansion, providing a kind of positive feedback loop.

Quantum theory predicts that virtual particles will appear from nothing and anihilate as vacuum energy

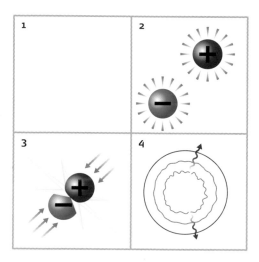

We're most familiar with positive feedback when a microphone gets too close to a loudspeaker. Any ambient noise is picked up by the microphone, amplified and comes back through the speakers. This noise adds to the background noise going into the microphone, so even more noise is pushed out by the speakers. In a small amount of time, the sound that is fed back on itself builds up to form the familiar screech of feedback.

Similarly, if the energy that is causing the universe to accelerate its expansion increases with the volume of the universe, then there will be more energy to drive that expansion more quickly, producing more energy from the expansion to drive the process even more quickly and so on.

For the moment, dark matter and dark energy both still have a degree of mystery about them. It certainly is amazing that everything we can see and detect out in the universe, every bit of matter in stars, planets and dust, every photon of light, all the energy in the nuclear reactions of stars, all added together only makes up four percent of what the rotation of galaxies and galactic clusters predicts should be out there.

Imagine only being aware of five percent of the population of a country. It is an incredible level of ignorance. We have the concepts of dark matter and dark energy – but dark matter may not exist at all, and dark energy is really only a description of what is happening and not a useful explanation for why it happens. There is much more to discover.

Intriguing though they are, dark matter and dark energy are not the only oddities out in space. Throughout the twentieth century, astronomers uncovered a whole zoo of strange objects in space, some of which are we only now beginning to understand. It's time to meet the cosmic oddities.

chapter 9

Cosmic oddities

There are plenty of remarkable phenomena in the universe beyond the friendly confines of the solar system. The basic building blocks of solar systems and galaxies may be stars, planets and dust, but there are hundreds of weird and wonderful components needed to build the complete universe that we see through our telescopes.

One of the first oddities to be discovered was briefly identified as a signal from an alien race. This discovery dates back to the early days of radio astronomy. In 1967, a researcher at Cambridge University, Jocelyn Bell, received a regular pulsing signal on a radio telescope. After eliminating the usual possibilities of local interference it became increasingly certain that this signal was coming from the stars. This steadily repeating pulse was given the code

The Crab Nebula, the debris of a supernova, first seen on Earth in 1054 AD

name LGM-1, where 'LGM' stood for 'little green men', a common phrase at the time for describing aliens.

The signal's discoverers later claimed that this name was only a joke, but it's hard to believe Bell and her supervisor, Anthony Hewish, didn't momentarily think that they had locked onto a message from an alien life-form. The pulses came every 1.3 seconds with startling regularity. It was like tuning into the ticking of a cosmic clock. Other than a transmitter, what else could produce such mechanical regularity?

After a long study, comparing radio signals with scans of the same area using optical telescopes, it was decided that the source was a special kind of star, which would be called a pulsar. Since LGM-1 many more pulsars have been found, and our understanding of them is almost certainly correct. A pulsar emits a powerful beam of light, in this case in the radio frequency. This beam pours out of the magnetic north and south poles of the star. But the star is also rotating at great speed, so these beams sweep around like the light from a lighthouse.

The LGM-1 pulsar was making a complete rotation in around 2.5 seconds (because there are two beams, the frequency of the pulse is half this) – frighteningly fast for something the size of a star. And as more pulsars were found it was realized that the first example turned out to be anything but the quickest. Some pulsars rotate in just 1/1000 of a second. The reason for the speed of spin is down to the nature of the star that forms a pulsar – they are neutron stars. (Not all neutron stars are pulsars, but all pulsars are neutron stars.)

A neutron star is a star that has collapsed. They are formed during a supernova, a vast stellar explosion that blows the outer layers of a massive star into space, leaving behind little more than neutrons, particles that are usually only found in the company of protons in the atomic nucleus. Without the repulsive force of the charged protons to keep them apart, the neutrons collapse into an incredibly dense mass. If you had a piece of neutron star about the size of a grape, it would weigh 100 million tonnes. What started off as a star about the size of our Sun would end up on the same sort of scale as the island of Manhattan.

Artist's impression of the sweeping beams of a pulsar

As all this matter compresses, any initial spin that the star had will be greatly magnified. As we saw with the formation of the solar system, the angular momentum, the oomph of the spin, stays at a constant level, like an ice skater pulling in her arms while spinning. The result of this shrinking is to increase the spin rate, and because neutron stars have shrunk by an enormous amount, the result is extremely fast rotations.

If you could ever approach a neutron star it would not be a comfortable experience. The nearest neutron star detected so far is either the pulsar J0108-1431, which is around 326 light years from Earth, or a more recently discovered star in Ursa Minor, nicknamed Calvera, which could be as close as 250 light years. But that's still

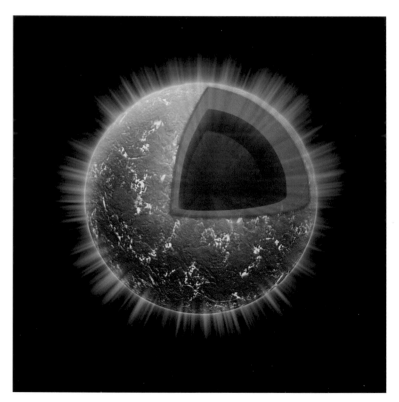

Cross section of a neutron star in Cassiopeia, around 11,000 light years away

The tidal force from a neutron star or black hole would spaghettify any visitors

quite a distance. Even so, if we ever developed the technology to cover that kind of distance you would discover that a neutron star system was not a friendly neighbourhood.

First of all, they are very hot. The surface of the Sun is about 5,500°C (the interior is much hotter, but it's the outside we experience directly). A neutron star, as the collapsed core of a star that went supernova, will typically have a surface temperature as high as 1,000,000°C.

Also, because you can get so much closer to a neutron star than you can to an ordinary star, because of its lack of bulk, the tidal forces that the star produces in you would be much bigger

Artist's impression of the impact of a black hole on its neighbourhood

than usual. Remember that the reason the Moon's gravity is bigger than you might expect is because you can get closer to its centre than we can the centre of the Earth. Get close to a neutron star and there will be huge differences in attractive force between one end of your spaceship and the other.

The result, as the nearer end of the ship is pulled away from the more distant end, is that your spaceship would be stretched into a long, thin strip in a process graphically known to astronomers as 'spaghettification.' And the same process would happen in you. You would be stretched into pink spaghetti. Neutron stars are the bad boys of the cosmic world. But even they seem timid when compared with our next, more famous, oddity of space, the black hole.

Leaving aside the physical nature of black holes, which we'll come back to in a moment, there is one big difference between a neutron star and its celebrity cousin. We know that neutron stars exist. We *think* black holes exist too – but the evidence, strong though it is, is all indirect.

Black holes have become part of the mythology of space and fictional space travel. In movies they are usually portrayed as totally black spheres in space with an irresistible pull, acting like vacuum cleaners that suck in everything around them. Get near a black hole and, Hollywood tells you, whatever you do you are going to be sucked in. The reality (assuming black holes do exist) is rather different.

Black holes are concepts that fall out of theory, in particular, from Einstein's general relativity. A version of the idea had been around considerably longer, though. Back in the eighteenth century, English astronomer John Michell wondered what would happen if you had a star that was so massive that its escape velocity was faster than the speed of light.

We all know that if you throw something up in the air, it will fall down again. 'What goes up comes down' is basic folk science. But if you throw something fast enough (or, to be precise, if Superman did, because we're talking a superhuman feat of strength) it could escape the Earth's gravitational pull and never return. This would mean throwing it at 11.2 kilometres per second or faster. You'll often read that this means that rockets need to reach this speed to leave the Earth's atmosphere, but that's just not true.

When a rocket is blasting away it is using the force of the motor to overcome the pull of gravity. The rocket can travel as slow as you like and still get into space. Escape velocity only comes into play for something like a thrown ball that has been given a big initial push but is then coasting with only gravity acting on it.

The heavier a body like a planet or star is, the higher the escape velocity. Michell thought that if a star was so massive that its escape velocity was greater than light speed then that star would

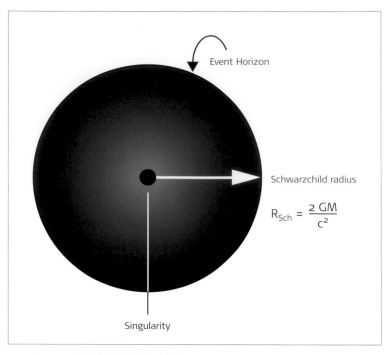

Event Horizon

Schwarzchild radius

$$R_{Sch} = \frac{2\,GM}{c^2}$$

Singularity

The structure of a black hole is simple but produces complex distortions in the material around it: the image on the right shows the colour shift from swirling gas around a black hole

turn black because the light wouldn't be travelling fast enough to escape. But this would only happen if light was influenced by gravity, which seemed unlikely, so no one took much notice until Einstein came along.

Einstein's general relativity showed that massive objects distort the spacetime around them. This is how gravity appears to work. The more concentrated the mass is, the bigger the curve it puts in the universe. German physicist Karl Schwarzschild realized, while in action during the First World War, that if you crammed enough mass into a small enough space, there would be so much distortion of spacetime that any light coming out of a body would be curved back into it. The light would never escape. Just like Michell's star, it would go dark.

This conceptual object, originally called a dark star or a black star, would later be given the name 'black hole' by American physicist John Wheeler in the 1960s. However, there was no suggestion that this really happened. Einstein explicitly said that he didn't think such stars could exist. After all, a black hole would be a lot denser than a neutron star. To get our Sun, a middle-sized star that is a lot bigger than the remnants that form a neutron star, compact enough for it to turn into a black hole it would have to be squashed down until it was just three kilometres across.

Yet the idea of black holes would not go away. Other astrophysicists came up with the surprising conclusion that there was a mechanism for black holes to form, a possibility that made it likely that black holes would exist. Stars are massive cosmic objects, with a huge gravitational pull dragging all the matter in them towards the centre. Usually that force is countered by all the energy generated by the nuclear reactions inside the star, puffing it up. But with a big enough star (the Sun is too small for this to happen), as the star becomes elderly and those nuclear reactions run out of energy, if the star doesn't explode, it is likely to collapse and form a black hole.

Although the Sun is too small to form a black hole under its own gravitational pull, this doesn't mean that all black holes would have to be incredibly massive. In principle, the matter in your body, for example, could be converted to form an incredibly tiny black hole. It's just that the atoms in it would have to be compressed with a force much bigger than anything we can imagine being applied to them – certainly much more massive than the force in a typical star.

So black holes could well exist. Although we don't have direct evidence of them, there are many observations that seem to show invisible black holes interacting with a nearby neighbour, where the outer parts of a star seem to be pulled off by something invisible but massive nearby. And even allowing for dark matter, the way galaxies rotate suggests that there are massive black holes, millions of times the mass of the Sun, sitting at the centre of each, including our own Milky Way galaxy.

Assuming that black holes do exist, they will have even more devastating tidal effects than a neutron star. They will spaghettify objects that come too near. But they won't be the cosmic vacuum cleaners they appear to be in Hollywood movies. The gravitational pull of a black hole will be exactly the same as the star that formed it. It will no more suck in everything around it than the Sun sucks in everything in the solar system. If you stay a decent distance away, it is just as easy to orbit a black hole, or to escape from it, as it is from any other star.

However, this doesn't mean that black holes don't behave strangely. The actual black hole is thought to be a singularity – the whole mass of the star is compressed into an infinitely small, infinitely dense point. Whether or not this actually happens is subject to much dispute. Not only have we never seen a black hole to check it out, our physical laws break down when infinity comes into the equation. However, we do know that a black hole would be incredibly tiny and dense. This singularity, if it exists, would be hidden away. As far as we're concerned what we would see (or rather wouldn't see) as a black hole is its event horizoThis is a spherical boundary around the singularity which is the point at which the gravitational pull is so strong that light bends in on itself and doesn't escape. When we describe the size of a black hole, what we really mean is the size of its event horizon. The event horizon is the point of no return. Anything – whether it is light or matter – that passes the event horizon is incapable of coming out again. Ever. Were it not for the faint glow of material from outside being sucked in and giving off radiation, the event horizon would be a perfectly black, perfectly uniform sphere, giving away none of the secrets of what lies beneath.

Seen from the outside, an object approaching the event horizon of a black hole would seem to get slower and slower. This is because general relativity tells us that the stronger a gravitational field is, the slower clocks run. This seems incredible, but the effect of gravity on time has a direct influence on a piece of technology many of us use to find our way around.

The Global Positioning System (GPS) or sat nav, compares the time and position broadcast by a range of satellites, using the time the signals take to get to the receiver to work out a position. There are two effects from relativity on those times. Special relativity, the version that deals with moving at different speeds, tells us that a clock moving with respect to an observer will be seen to run slowly by that observer. The clocks on the GPS satellites lose about seven millionths of a second every day because of it.

But general relativity also influences those satellites. The gravitational pull is less on the satellites than it is on the Earth, because the satellites are further away from the centre of the Earth. Where

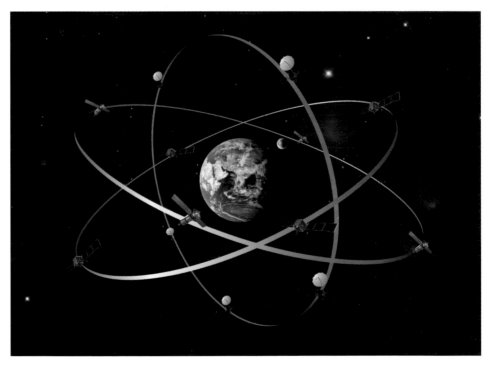

GPS satellites depend on clocks that are influenced by relativity

there's less gravity compared with an external observer, clocks run faster according to that observer. So, seen from the Earth, the clocks on the satellites gain 45 millionths of a second each day. Combining the two factors mean the clocks gain around 38 millionths of a second daily. It doesn't sound much, but if GPS didn't correct for the influence of gravity on time, the system would get your position wrong by several kilometres in just one day.

As something gets closer to a black hole and the gravitational pull increases, time slows down as seen by an observer at a safe distance. As the object reaches the event horizon of a black hole, still as seen from the outside away from the black hole's gravitational field, an object would get so slow that it would take an infinite amount of time to cross it. But in practice that observer would never witness this happening. As an object gets closer to the horizon, the light coming away from it is red shifted. By the time the object gets to the horizon, the light will have shifted all the way down the spectrum. There will be nothing detectable coming from it – not just no visible light but nothing electromagnetic at all. It would disappear.

Things would seem totally different if you were on a ship heading towards a black hole. If you could get past the horizon without the gravitational effects pulling you apart you wouldn't notice the event horizon passing. (You won't necessarily get spaghettified before crossing the horizon. Strangely, the bigger the black hole, the gentler the tidal force at the horizon.) There would be no obvious change as you passed the point of no return. The only thing would be that if you decided to turn around and get out, it would be too late. There would be no escaping being dragged into the singularity, bombarded by the ultra-high energy particles flying in with you, and shredded by the intense tidal force.

You might think the one thing we know for certain about black holes is that they are black. They certainly don't allow any light out and nothing can escape across the event horizon. And yet Stephen Hawking, the greatest living expert on black holes, has predicted that they should give off a faint glow. This is because, as we've already seen, 'empty' space is actually full of pairs of virtual particles – one matter, one antimatter – that momentarily pop into existence, then disappear again without ever interacting with anything.

When these virtual particles appear near a black hole's event horizon, one particle will occasionally get sucked into the black hole while the other flies off in the opposite direction and escapes. The result is that radiation emerges from just outside the event horizon. A black hole should give off the faint glow of that radiation.

The energy for this radiation has to come from somewhere – and the place it comes from is the black hole. Over time a black hole will lose energy and contract. This means that if very small black holes are created, perhaps by the high energy collisions that might be possible at the Large Hadron Collider at CERN in Switzerland, the micro black holes will almost instantly disappear as they lose energy through this 'Hawking radiation' and shrink to nothing.

Another way it's thought that black holes could be made visible is if a black hole is in a binary system with another star. It's quite common for a pair of stars to be orbiting each other in a single solar system. Remember how the solar system formed from a rotating disc of material, pulled together by gravity. If there had been two major points of concentration, where the matter built up quickly, it would be easy to end up with not one but two stars in a system – and this often happens. (There are even systems with more stars in a complex relationship.)

If one of the stars in a binary system is a lot more massive than the other, its extra gravitational pull tends to suck material off the companion. Should the massive star become a black hole, this would continue to happen, and we would see that material heading off into nothing. (Note again, the black hole is no better at stripping its fellow than the pre-collapsed star.) There have been a number of observations where material from a star appears to be spiralling into an invisible partner, strengthening the case for the existence of black holes.

In the movies, black holes are often treated as if they were portals through space and time, providing a way of getting to somewhere else much faster than the speed of light. Although, as we'll

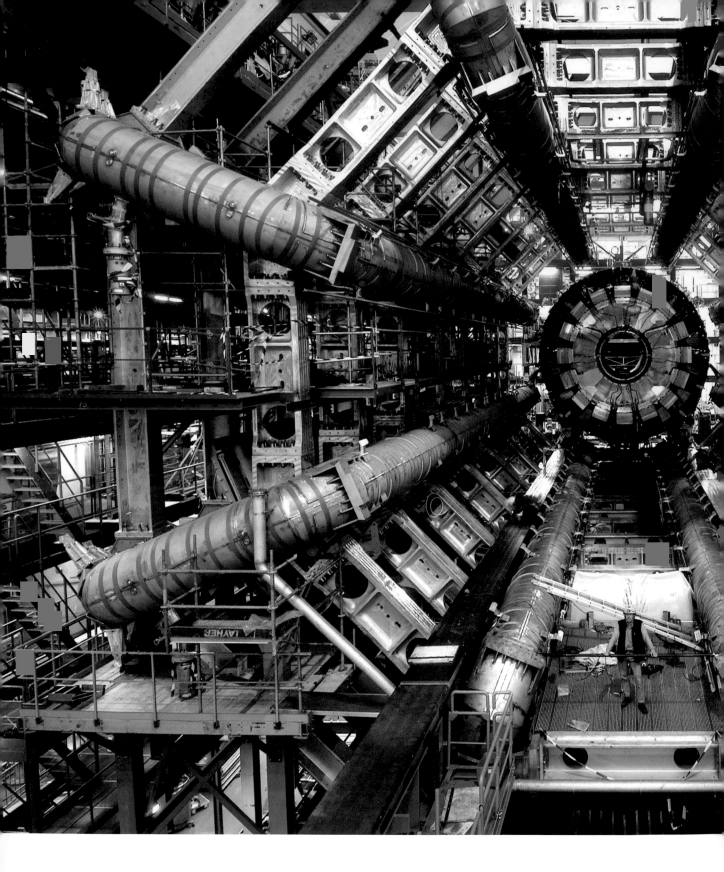

The ATLAS detector in the Large Hadron Collider at CERN

Artist's impression of a black hole pulling material from a binary star

discover, this wouldn't work, there is an underlying bit of sense that reflects the way that general relativity, the theory that predicts the existence of black holes, treats mass and gravity. In general relativity, mass is shown to distort spacetime. The more concentrated the mass, the more spacetime is warped by its presence.

If you imagine spacetime as being a large sheet of rubber (remember it is actually four dimensional, but we are considering just two dimensions for ease of imagining), then the effect of the mass of a star would be to cause a big indentation in that sheet. If you concentrate the whole mass of that star at a point, you would get a very thin but incredibly long indentation in the rubber, like an infinitely long cone getting thinner and thinner but never quite reaching zero width. Although a black hole appears to be the size of its event horizon, spacetime is so distorted near the singularity that it heads off infinitely far, effectively in a different dimension.

The idea of a black hole acting as a portal is that if, somehow, you could pass through the singularity, down that funnel in space-time, you would end up coming out somewhere else. But this would not be a manageable 'somewhere else' by a nearby star, or even in a nearby galaxy. It would be a somewhere else in a whole new universe. The distortion in spacetime caused by the singularity would take you outside our universe, but the point is academic. A singularity is totally destructive. Nothing escapes. You can't pass through it.

Remarkably, though, there may be a way around this, but it depends on the black hole spinning. It seems highly likely that

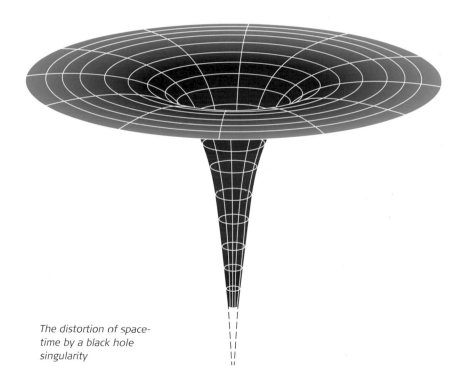

The distortion of space-time by a black hole singularity

this would be the case. Everything else seems to spin, and, like a neutron star but more so, a black hole should spin much faster than the original star after it has contracted. This makes it possible that the singularity could be spun into a circle with an empty space inside it, so you could shoot through the middle and survive the journey. This is fine – but you still couldn't get out of the black hole. Nothing gets out.

You may have heard about 'wormholes' being suggested as the way to get through. A wormhole is one step further than a black hole. Here, rather than spacetime being infinitely stretched into a point by a singularity, a wormhole is a tear in the fabric of spacetime. If you think of our rubber sheet again, imagine that sheet being folded into a U shape. To get from the top left of the 'U' to the top right you have to travel through spacetime all the way around the letter. But a wormhole would be a tunnel linking the two top parts of the U – you could get from one place to the other without going through all the space in between.

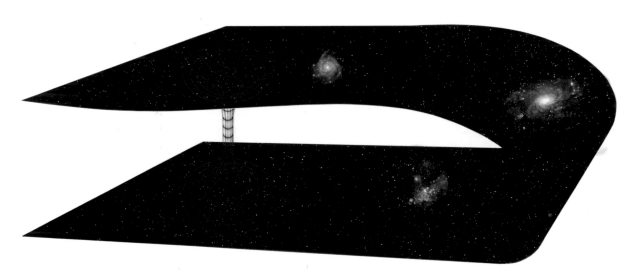

A wormhole providing a tunnel between two distant points in space

Very small wormholes, which are like tiny tubes connecting two separate points in spacetime, may be produced naturally, but they could never be traversed. Not only would they be too small, but they would collapse if anything tried to pass through them. (I need to stress no one has ever seen a wormhole. They are much more hypothetical than black holes.) Much larger wormholes have been envisaged as forming from a pair of white holes, or a black hole and a white hole – but there are problems with this, number one being that white holes probably don't exist.

A white hole is an anti-black hole. It's rather like a small-scale version of the Big Bang. It is based on a time-reversed singularity, which instead of sucking everything in like a black hole, pushes everything out. There's no problem getting out of a white hole – in fact you can't avoid it. But, with the exception of the Big Bang, physics seems to predict that white holes can't exist, which is something of a problem. Certainly, unlike black holes, there has been no evidence for them discovered yet.

Black holes are a special inhabitant of the universe that we can't see – but there are other strange phenomena out there that are all too visible. Sometimes an existing star suddenly bursts into extra brightness. In the early days, before telescopes, these stars would seem to appear out of nowhere and would be labelled as new stars – quite literally. The Latin word for new is 'nova' and that's what they were originally called.

One of the best-known novae burst into the night sky in 1054. During that year, wherever the constellation of Taurus was visible, there was a new bright star that hadn't been seen before, so bright that it could be seen in daylight for nearly a month before it gradually faded away. Now classified as SN1054, the fuzzy remnants of this nova form what is now called the Crab Nebula.

As we've come to understand what happens when a new star like this appears in the sky, the nova has been divided into two quite different phenomena – a nova or a supernova. A nova is a bit like a runaway version of the black hole that sucks material from a companion star. In a nova it is a very bright, small star, a white dwarf, which drags in matter from a larger star nearby.

Unlike the slow process in a normal star, all the new hydrogen pulled off the companion immediately gets compressed and heated in the white dwarf. The result is a massive explosion that blows away the outer parts of the white dwarf, which we see forming a nova. This process can be repeated many times. But it is nowhere near powerful enough to explain the degree of brightness seen in 1054.

The explosion that formed the Crab Nebula is now classified as a supernova. These are on a wholly different scale. They can start off the same way as a nova, in which case much more of the white dwarf and its accumulated matter undergo a fusion reaction all at once. Or, alternatively, the whole star could reach the point of no return, with its inner parts collapsing to form a neutron star (see page 158) and all the outer matter blasting off in a huge explosion. In a few weeks, a supernova can produce as much energy as the Sun will pump out in its entire lifetime. In its brief moment of glory a supernova can make a single star almost as bright as a whole galaxy.

Yet bright though they can be, supernovae don't get the crown as the most intense objects in the universe. This award goes to the mysterious phenomena called quasars. Quasars look like stars – hence the name, which is a contraction of 'quasi-stellar object.' Yet if they really were stars they would be impossibly bright. We see quasars at the limits of observation, vastly distant and, because of the expansion of the universe, heavily shifted towards the red end of the spectrum.

Because we are looking a long way into space, the light has been travelling for aeons. Peering this far back we are seeing into a distant past when galaxies were young, and it is thought that quasars are the core of early galaxies. You could call them baby galaxies.

Most, and quite possibly all, galaxies appear to have a super-massive black hole at their heart hundreds of times the mass of the Sun. A quasar seems to be the effect of this newly formed black hole sweeping up debris from the area around it, for once acting like they do in the movies and sucking in everything within reach. By the time a galaxy matures, this whole area is empty with nothing left to absorb and the black hole merely acts as a sort of galactic counterweight. But in those early days there was plenty of matter within reach, and as that matter is accelerated into the black hole it gives off radiation in the form of high energy light.

This can make quasars as bright as a typical galaxy, producing a single object that is giving off as much light as a trillion stars. Although these remarkable phenomena have only been known about since the 1960s, when they were first discovered by radio telescopes, we have now catalogued over 200,000 quasars. These remarkable intergalactic lighthouses have been discovered at distances varying from around 2.5 billion light years to a massive 28 billion light years, although they are much more common at greater distances (and hence earlier times).

To begin with there was much argument about just how far away quasars were. They were *so* bright that it seemed ridiculous that they could be billions of light years away, so there was some concern that the whole standard candle approach (see page XX) of measuring distances in space had gone wrong. It was only in the 1980s that astronomers and astrophysicists agreed that quasars were these far-distant baby galaxies.

Before settling on the baby galaxy theory, several theories had jostled for attention. For example, astronomers who did not believe they could be so bright assumed that they were closer than they appeared to be. This meant explaining the red shift a different way. It was thought that they perhaps they had such a huge red shift because an immensely powerful gravitational pull was dragging the light down the spectrum.

In the end, as we discover time and again, astronomers and astrophysicists are limited in what they can do when compared with other scientists. They can't go out and see what's going on in the depths of the universe, nor will they ever be able to do so. They can't experiment on a star in the laboratory. This means that most of their theories about the nature of the distant parts of the universe, particularly those governing cosmology and the universe as a whole, have a higher degree of uncertainty than those applying to, say, conventional physics.

Black holes, for example, probably do exist – but there are alternative theories to explain their effects. Dark matter is even more tenuous, with some evidence suggesting alternative theories may be better. We think we know what quasars are, but we can't be a hundred percent certain. Even something as fundamental and as widely acknowledged as the Big Bang is, in the end, speculative.

Although astronomers and cosmologists tend to speak about the Big Bang as if it were fact (especially when they're let loose on TV shows), it isn't anywhere near as solid. All we can say is that the Big Bang is the best-supported theory, and that the evidence uncovered to date fits well with the modified Big Bang theory (the original version, for example, didn't have inflation, which means it didn't match observations). So there's no harm in treating the Big Bang as if it were a pretty solid idea, because it's the best we have. But there are some intriguing alternatives to explain the early life of our universe.

Artist's impression of the jets produced as matter streams into a quasar

chapter 10

Just a theory

In the early days of its existence, midway through the twentieth century, the Big Bang theory was one of two competing explanations for the way the universe developed. The other, called the Steady State theory, had a lot going for it. The idea behind Steady State theory is that the universe is (and always has been) growing at a constant rate. According to this theory, matter is gradually but constantly coming into existence throughout the universe. If this is the case the universe had no beginning and will have no end, maintaining a constant flow of being.

Steady State was a neat theory because it didn't have a lot of the difficulties facing the Big Bang. On its own, the Big Bang has trouble explaining why the universe is, on average, uniform in every

The Hubble ultra deep field, showing some galaxies as they were less than 1 billion years after the Big Bang

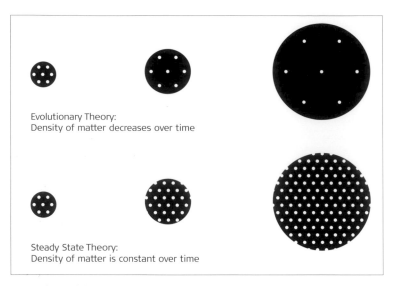

Evolutionary Theory:
Density of matter decreases over time

Steady State Theory:
Density of matter is constant over time

Comparison of the evolutionary Big Bang and Steady State theories

direction. It might seem that the universe is lumpy – if you look into space there's a lot of variation between galaxies and empty space – but if you take the universe as a whole it is relatively evenly spread.

The concept of inflation was partly introduced to cope with this otherwise difficult-to-explain uniformity of the universe, as its incredibly fast expansion would spread similar conditions across a large volume of space. However, for the Steady State theory, having a uniform universe is a natural state of affairs. And because a Steady State universe has no beginning there is no problem with explaining why the universe came into existence in the first place – it has always been there.

Of course it was perfectly reasonable to ask the Steady State supporters, notably the eccentric genius Fred Hoyle, 'Where is the new matter coming from?' but Steady State didn't have an answer for this. Its originators simply assumed that there was an unknown field that was responsible for bringing the matter into existence. But, to be fair, the Big Bang theory has arguably an even bigger problem. Big Bang doesn't answer 'Where did the singularity that formed the universe come from?' so Steady State was no worse in this respect.

You might also wonder, if Steady State was true and matter

was constantly popping into existence, why we don't see this stuff appearing from nowhere all over the place. Fred Hoyle, in answer to this, pointed out that it would only require one atom to come into existence once a century in a volume of space the size of the Empire State Building to maintain the expansion of the universe. In any particular patch of space, it simply wouldn't be noticed.

It was Hoyle who gave the alternative theory the nickname 'the Big Bang' in a radio broadcast comparing the two theories, and for a while it did look as if Hoyle's theory would push Big Bang aside. But with information about the early universe flowing in from the new radio telescopes, Hoyle and his colleagues got a shock. If Steady State were correct, then the universe should look pretty much the same, however far you looked into the past. This was, after all, not an evolving universe but a constant one. Unfortunately for Hoyle, all the evidence showed that there were a lot more young galaxies then than there are now.

The upshot was that the Steady State theory was abandoned. This wasn't a scientific decision per se, but more a matter of following the herd. Steady State was the idea of a relatively small group, whereas Big Bang was the work of a host of US universities. The Steady State theory could have been modified to match the new data, just as the Big Bang theory was changed to fit contradictory data, and Hoyle demonstrated as late as 2000 that a modified Steady State theory could work with current data, including the cosmic microwave background – but by then no one at the forefront of cosmology and astrophysics was listening.

So it's still possible that a modified version of the Steady State theory is better than the Big Bang – but there are many more possibilities, some much more exotic. One is that the Big Bang existed, but that it didn't start with the beginning of all of space and time in a singularity. Some cosmologists believe that there was a universe before the Big Bang and that this universe went through a process where it became very small, contracting until it bounced, but never quite reaching a singularity.

This is attractive both because it does away with the infinities of the singularity and it explains where the Big Bang came from. But we know that the universe can't simply be going through a cycle of contraction and expansion. This is in part because its acceleration means it may never contract, but also the laws of thermodynamics suggest that such a cycle would be like a perpetual motion machine – attractive in concept but impossible to realize. However, a new theory that attempts to explain how gravity works, called loop quantum gravity, does make it possible for such a 'Big Bounce' to happen, and in principle there should be remnants of the pre-bounce universe to show whether or not this was the case.

Returning to the conventional Big Bang, others give it a more exotic context. This reduces our universe to just a tiny portion of the whole of existence. This is confusing as 'universe' should mean everything. The most accurate description is probably to still use the term 'universe' as everything, and describe our part of it as a 'pocket universe', but it is more popular to keep 'universe' as a term for the bit we know best and to refer to the larger whole as a multiverse.

In this picture there are many parts of the multiverse, perhaps even an infinite set of them. Anywhere within the multiverse, many times and in many places, Big Bangs can take place. This results in a tiny segment of the original multiverse blowing up like an expanding soap bubble that is part of a whole froth of bubbles. So in this picture the true universe could go on for ever in time and space, with just our local segment experiencing the beginning of our local spacetime in the Big Bang.

In such a multiverse, each separate universe could have quite different natural laws. In many of them, matter may not even be stable – there could be plenty of empty universes. Or, equally, a universe could be almost identical to our own. We are unlikely ever to know. Although physicist Michio Kaku has suggested that if a civilization survived for billions of years it might develop the technology to travel from one universe to another, it would require the sort of trickery with black holes, white holes and wormholes that we've already seen, which is far beyond anything we can imagine now.

Other scientists have come up with a possible description of

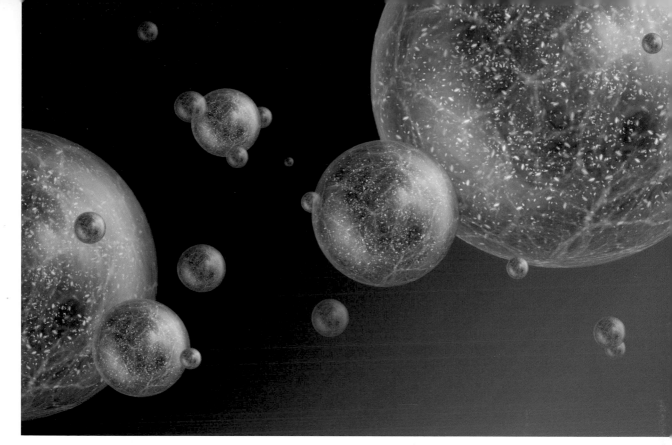

Our universe could be a bubble in a larger multiverse of Big Bangs

the universe that is even stranger than the multiverse. To understand it, we need to take a quick excursion into the wonderful world of string theory. This is not about bits of string in the normal sense, but is a possible model for the nature of the forces and particles that make up our universe.

It's quite possible that the biggest and most potentially rewarding problem in all of science is finding an explanation that works consistently for all the different forces that control the universe. Three of these forces seem to fit together fairly well. But the science we have to describe the fourth, gravity, doesn't work with the others. Similarly there are two physical theories that cover pretty well everything at the fundamental level. Quantum theory covers the very small, and general relativity describes the nature of spacetime and how large things behave. Unfortunately these two theories are incompatible. They just don't fit together.

What's needed is a theory of quantum gravity, one that deals with gravity (which means general relativity), but incorporates the nature of quantum theory. This would enable scientists to pull gravity in with the other three forces of nature. There are a number of approaches to doing this, including the loop quantum gravity we have already met. This theory is gaining support, but, at the moment, the most worked-on method for combining gravity and quantum theory is a rival approach called string theory.

In a basic description string theory sounds both elegant and pleasingly simple. All you do is replace all the particles – both particles of mass like electrons and quarks and the particles responsible for forces like photons and the gravitons that are thought to carry gravity – with a single type of object called a string. This is an unbelievably tiny one dimensional entity that can vibrate in a whole host of different ways – and the idea of string theory is that the different vibrations in a string produce all the zoo of particles that make up reality.

Although this description sounds simple – and feels as if it makes a kind of sense – the reality of the maths behind the picture is fiendishly complex. To make it work, the universe has to have ten dimensions of space and time, rather than the familiar four. And as yet string theory makes no testable predictions, so it is impossible to disprove, making some argue that it isn't really a scientific theory, just an abstract tour de force of mathematics. However, if string theory is along the right lines, it opens up the possibility of a strange mechanism for the universe to produce an effect like the Big Bang.

This is described in the strangely named Ekpyrotic theory. This approach imagines the universe as a four-dimensional membrane floating in the multidimensional space required by string theory. (Technically this approach uses M-theory, which is a more advanced version of string theory incorporating an extra dimension.) In this picture, the Big Bang is how we experience a collision between two such membranes, one of which is our universe. When this happens our membrane, the universe as we know it, is wiped and begins expanding as it bounces away from the collision. And unlike the

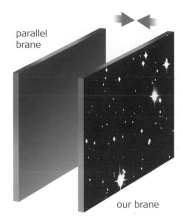

parallel brane

our brane

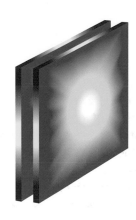

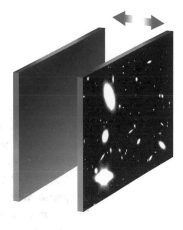

Two branes are attracted towards each other – one will form our universe.

As the branes collide, the energy of collision converts to radiation and energy in the big bang.

The branes travel away from each other, expanding. Gravity pulls matter together to form stars and galaxies.

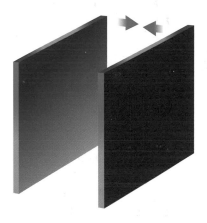

The branes slow down. As our universe expands, matter thins out in it.

Attraction overcomes the movement of the branes and they reverse. Expansion accelerates in our universe.

In the Ekpyrotic universe a pair of branes collide to produce the Big Bang

traditional Big Bang this can happen over and over again, with the universe expanding every time. It never stops expanding.

The Ekpyrotic theory conveniently does away with the need for dark matter and dark energy which are effects produced by the nature of the membrane universe. It also explains why gravity is so

Some cosmologists speculate that the universe could be a Matrix-like simulation run on an alien computer

weak compared with the other forces in nature. (Remember the comb and the pieces of paper.)

In Ekpyrotic theory, it's assumed that gravity leaks out of the familiar dimensions in the membrane into the wider multidimensional space around it, drawing the two membrane universes together to make them collide. If this is the case, much of the true gravitational force would never be experienced in our universe and it would seem weak – as we know it is.

There are many other strange and wonderful forms the universe could take. One, for example, puts the universe inside a black hole, while another suggests that it exists in the form of a hologram with fewer dimensions than it appears to have. But one last concept that's worth considering is that the universe we experience is a program running on a computer in the 'true' universe. This would put us all in a vast equivalent of the movie *The Matrix*, although we would all be computer-created constructs, not 'real' people projected into the matrix.

The idea is a little like the computer game *The Sims* but run on the scale of a universe. Imagine that there were alien races so technologically advanced that they could run a whole simulated universe on their computers. (If this sounds overly complex, the whole universe wouldn't need to be filled out in detail. It's a bit like the movie *The Truman Show*. The parts we experience would have to be detailed, but the rest could just be sketched in as required.) It's entirely possible that the universe we experience is just such a simulation, whether run for scientific purposes or purely for entertainment.

Like string theory, and several alternative models of the universe, this idea of living in a computer program suffers from being un-testable. Of course, if the programmer decided to set up a row of galaxies to spell out 'Hello world' or there were other obvious modifications – or computer bugs – the concept might become testable, but for the moment we have to leave it as a fascinating but not particularly useful alternative. And to hope that, if it is true, they don't turn us off any time soon.

If the universe really were running on a computer it would, inevitably, come to an end at some point. But stepping back from the idea of a simulation and returning to the standard Big Bang model, what is the future for our universe? Will it end with a bang, a whimper, or not at all?

chapter 11

The ultimate question

There are few experiences more enjoyable than looking at the sky on a dark, starry night, particularly if you are well away from civilization where there is no pollution and there seem to be many more stars in the jet black curtain of space. It feels like we are peering into the depths of the universe.

We now understand the scale of that vista. It's not just a few planets and a sphere of stars circling around the Sun, but a vast array of unbelievable grandeur with galaxies in their billions, each occupied by countless stars. It seems like a structure that is beyond the ravages of time. But are we looking at something that will last forever? What is the fate of the universe?

Artist's impression of a dead planet crossing the face of a red giant star, seen from an alien moon

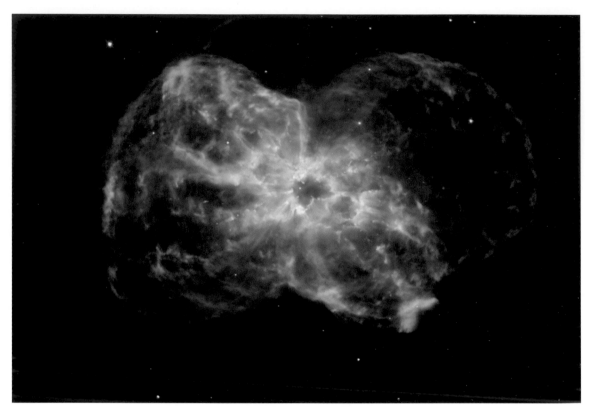

A sun-like star casts off its outer layers of gas to form a white dwarf

Using the simple fact that the universe is expanding, and seems to have been doing so for a very long time, we were able to work out that there was a beginning, or at least a crucial change point that we call the Big Bang, which happened around 13.7 billion years ago. But things aren't as clear when we look forwards.

Before it was discovered that the expansion of the universe was accelerating, there was a lot of speculation that this growth might run out of steam. After all, there's lots of matter in the universe. Each atom of stuff is attracting all the rest. Just as a ball thrown in the air eventually comes to a halt and falls back under the influence of gravity, so it was thought that the universe might slow to a stop before beginning to contract. Eventually, after living in reverse for as long as it took to get to that turn-round point, it was thought that the universe would collapse into a single entity – sometimes called the Big Crunch.

Only, as we have seen, the expansion isn't slowing down. The universe is speeding up its stretch into nothing. So where will it end up? Will it be spread infinitely thin? Let's start with our own back yard. We know there was a time when the Earth didn't exist. What will happen to the Earth in the distant future?

Our Sun has been around for about 4.5 billion years. It should continue in a fairly similar form for another five billion years or so, by which time it will have burned through most of its hydrogen fuel. As it switches to fusing helium, the element produced by its hydrogen fusion, into still heavier ions, the Sun will expand, but its larger and cooler (and hence redder) surface will hide a much hotter core. In as little as a billion years after it forms, this red giant will also have become unstable and will blast off its outer layers, leaving a relatively tiny white dwarf star behind. (Only in cosmology can you say 'as little as a billion years'.)

Sadly, our descendants will not experience this remarkable transformation – unless they set up an observation platform at a safe distance. When the Sun expands to form a red giant, the chances are that the Earth will be burned up. (This is not absolutely certain, however. The Sun will have less mass by then so the Earth's orbit will be bigger, making it possible that it will avoid disappearing into the red giant.)

This potential fate demonstrates just how big the red giant will be. The Earth is around 150 million kilometres from the Sun – it takes sunlight, even at its blistering speed of 300,000 kilometres a second, eight minutes to reach us – so the red giant is expected to be over a hundred times bigger than our Sun is now.

We won't be able to live on Earth that long though. During the period before it turns into a red giant, the Sun will be heating up, a process that has been going on throughout its life. Around a billion years from now, the Earth will be uninhabitable. If our descendants

A cluster of distant galaxies, some distorted by gravitational lensing

are still around, they will need to have evacuated before then. But this shouldn't present us with too much of a problem.

A billion years is around ten thousand times longer than human beings have been in existence. Bearing in mind that we've only had aeroplanes for around a hundred years, and space travel for just over half that, a billion years gives us enough time to find a way to escape. We have more pressing problems than the gradual warming of the Sun.

If there is any way to travel interstellar distances, and the human race (in some form) survives, it seems likely that we will have cracked the problem before the Earth becomes uninhabitable. Should we return to watch the Earth's demise, we may also see the whole solar system torn apart. At around the same time that the Sun expands to form a red giant, the Andromeda galaxy is expected to collide with the Milky Way.

Although this will be a meeting of two immensely large bodies, they are mostly composed of empty space, so we wouldn't expect a single vast collision. Some stars will be captured by others. Some bodies will collide. But the future of the solar system cannot be projected. Our Sun, relatively near the edge of our galaxy, could be flung off into intergalactic isolation or it could remain with one or other of the two galaxies' main clusters of stars.

Even if the human race does survive this long and has developed the ability to travel around the universe at will, this doesn't mean that we can simply jump from one habitable planet to the next for ever. This is because it isn't just stars that run out of oomph over time. What we expect to happen is that the whole universe will run down like a clockwork mechanism that hasn't been wound since the Big Bang.

As the expansion continues over many billions of years, as hydrogen burns to helium and helium to heavier elements, unless there is a catastrophic change in the universe – another Big Bang, perhaps – we will end up with less and less energy in any particular section of space. The final outcome is liable to be cold, empty space with just isolated bits of matter, each having so little energy that they hardly move.

Quantum theory, the fundamental theory behind the existence of matter, says that particles can never come to an absolute stand-still. The result, rather, would be a little like the series 1, 1/2, 1/4, 1/8, 1/16... Each time a fraction is added to the series it is smaller than the previous one, but it will never reach zero. Similarly, the particles that make up the universe will have less and less energy, but they will never finally come to a stop, never reach a zero energy state. Eventually, though, however many billion years it takes, there will be insufficient energy for any conceivable form of life to exist.

It sounds depressing, this universe of the far future, a dull place where nothing much happens and there is nothing to see. (Nor, for that matter, will there be anything or anyone in the universe to see what remains.) But we have to realize that we are talking about a timescale that exceeds the entire lifetime of the universe so far. Modern human beings have existed for about 200,000 years, civilizations for maybe 10,000 years. From our viewpoint, the sort of timescale we're talking about until we reach a cold, lifeless universe is unimaginably long. For any foreseeable future, the universe will remain a wonderful, fascinating place.

There will still be stars and planets. There will still be the vast, spinning arrays that make up the galaxies and the weird and wonderful cosmic zoo that is occupied by black holes, supernovae and quasars. The night sky will still be a thing of wonder. Even if we explain dark matter and dark energy, there will be many mysteries for science to ponder. And as long as there is humanity to look, we will peer into the darkness, and marvel at the sheer scale of the universe.

We may have begun to explore what is out there, but space remains that final frontier – and within any humanly imaginable timescale it will be a frontier of never-ending discovery.

Picture credits

Every effort has been made to credit the appropriate source. Please contact us if you find any errors or omissions so that these may be corrected in any future printings.

Courtesy of nasaimages.org: NASA/STScI **1, 141**; NASA **2**; NASA/ESA and G. Bacon(STScI) **6–7**; NASA/JPL **10**; NASA/JPL **18**; NASA/Dana Berry **40–41**; NASA/JPL **48**; ESA/NASA/SOHO **50**; NASA/JPL **54**; NASA/JPL/USGS **57**; NASA/JPL **58–59;** NASA/GSFC/Arizona State University **61**; NASA/ESA, The Hubble Heritage Team (STScI/AURA), J. Bell (Cornell University) and M. Wolff (Space Science Institute, Boulder) **66**; NASA/JPL **67**; NASA/GSFC **68–69**; NASA/JPL/University of Arizona **73**; NASA Planetary Photojournal **74;** NASA/JPL/University of Arizona **76**; NASA and Hubble Heritage Team (STScI/AURA) acknowledge: R.G. French (Wellesley College), J. Cuzzi (NASA/Ames), L. Doves (SwR), J. Lissauer (NASA/Ames) **78–79**; NASA/JPL/Space Science Institute **80**; NASA/JPL/USGS **81**; NASA/JPL **83**; National Sciences Foundation **86**; NASA/JPL-Caltech **87**; NASA/JPL-Caltech/ STScI **88**; NASA, NOAO, ESA, The Hubble Helix Nebula Team, M. Meixner (STScI), T.A. Rector (NRAO) **95**; NASA/JPL-Caltech **100**; NASA **102**; NASA/JPL-Caltech/ NRL/GSFC **117**; NASA/CXC; UV; NASA/JPL-Caltech. Optical NASA/ ESA/ STScI/AURA, IR: NASA/JPL-Caltech/University of Arizona **120–121**; NASA/JPL-Caltech/Harvard-Smithsonian CFA **122**; NASA **124, 127**; R. Svobada and K. Gordan/NASA **129**; NASA/WMAP **134–135**; NASA, H. Ford (JHU), G. Illingworth (UCSU/LO), M. CLampin (STScI), G. Hartig (STScI), the ACS Science Team, and ESA **144**; NASA, ESA, M. Livio (STScI) and the Hubble Heritage Team (STScI/AURA) **146;** NASA/ ESA/JPL/Arizona State University **156**; NASA/JPL-Caltech **162**; NASA, ESA and N. Pirzkal (STScI/ESA) **180**; NASA/JPL/STScI/AURA **192;** NASA/ ESA/JPL-Caltech/STScI **194.**

Science Photo Library: Pasieka **14**; Jean-Lou Charmet **16**; Sheila Terry **29, 31;** Gianni Tortoli **30**; Mark Garlick **32, 84–85;** GIPPhotoStock **36**; Science Photo Library **47, 133**; Rev. Ronald Royer **62**; WMAP Science Team **105**; Science Source **110**; NRAO/AUI/NSF **142**; R. Massey, Caltech/ NASA/ESA/STScI **149**; Lynette Cook **150–151, 178, 190**; NASA **159**; NASA/ CXC/M. Weiss **160**; NASA/ESA/STScI/G. Bower and R. Green, NOAO **165**; Julian Baum **172**; Victor De Schwanberg **174**; NASA **185.**

Other sources: SV2 Studios **9, 112;** Courtesy of U.S. Air Force **12**; photograph Golden Meadows **15**; Snapler at Pixishots **44**; Courtesy of John Reed **91**; Courtesy of ESO/H.H. Heyer **92–93**; Courtest of Wabash Instruments Corporation (1961) **101**; Don Dixon/Cosmographica **108**; Courtesy of Brookhaven National Laboratory **114;** Courtesy of W.M. Keck Observatory **126;** Courtesy of Kamiola Observatory **130–131**; Courtesy of Caltech/Palomar Observatory **136–137**; Courtesy of F. Millour OCA, Nice, France **136–137;** Courtesy of NAIC-Arecibo Observatory, a facility of NSF **136-137**; Courtesy of Maximilien Brice, CERN **170.**

Index

Note: page numbers in *italics* refer to illustrations